~ RALSTON'S GOLD ~

GEORGIA, COLORADO, CALIFORNIA

GOLD STRIKE PARK

To Earl and Jean Almquist

What wonderful friends and supporters you are!

Lois Lindstrom

June 22, 2012

Lois Lindstrom

Ralston's Gold

First Edition
Copyright ©2011

Inquiries should be addressed to:
Lois Lindstrom
6550 Yank Way #211
Arvada, CO 80004

Coloradream Publishing

Edited by Lisa Langley

Designed by David Robison
www.Coloradream.com
First printing: November 2011

Printed in the United States of America

Dedicated to my family:
Linnea and Tod
Perrick Robert
Karl Wallace

Contents

❧ FORWARD ❧

Lewis Ralston found gold in a stream named that day for him on June 22, 1850. The stream flowed through present-day Arvada, Colorado, but in 1850 the site was part of Kansas Territory - Colorado and Arvada had not been born.

Ralston was part of an 1850 group pioneering a trail through virgin forest and plain to reach the gold discoveries in California. The fur trapper's forts, erected in 1830 along the Platte River, were, by 1850, heaps of adobe.

Bent's Fort, a magnificent adobe palace built by William Bent, was in ruins. Bent had destroyed it himself when the then federal government refused to pay for its use. Fort Pueblo, with broken walls and a sagging gate, was the last of the fur forts. It was razed by Ute Indians in 1854. Fort Massachusetts, a federal fort, was built in 1852, the same year that Mexican sheepherders founded San Luis, the oldest town in Colorado.

Two major roads touched the area that would become Colorado. The Oregon Trail, opened in 1811, nipped the northeast corner, and the Santa Fe Trail, broken by traders in 1821, left the Arkansas River and bent south along the Purgatoire River. Both avoided the towering barrier of the Rocky Mountains.

The Arkansas River was the northern border of Mexico. After the Mexican War (1846-1848), part of the area was claimed by New Mexico, but statehood was not secured until 1912.

In 1850 there were no settlements or outposts of civilization in the area that would become Colorado. There had been earlier rumors and claims of gold discoveries. None of these had a location or an exact date, and none of these finds brought an exploring gold-seeking party that would lead to the development of a state.

The Ralston gold site is the first documented find in the area that had been protected for centuries by the majestic Rocky Mountains. Lewis Ralston's name and his discovery were all but lost to history. The people who returned to his site founded Colorado. *Ralston's Gold* tells his story.

CHAPTER I

～⊙ GOLD ⊙～

Gold, treasured since the beginning of time for its beauty, was also a symbol of power because it was so rare. Only royalty and wealthy individuals could afford to adorn themselves in gold, and only prosperous merchants used gold as a medium of exchange. Gold flakes and nuggets were difficult to evaluate and a system of coinage was developed.

More than 3,000 years ago, King Cyges of Syria (Turkey today) ordered his goldsmiths to produce the world's first gold coins. These coins, to be trusted, had to be pure gold and the standard of measurement was a karat. The name came from the pod of the carob tree because each carob pod weighs one-fifth of a gram. The term "24 karat" indicates pure gold. [1]

The term grain indicates the weight of gold, named after an individual barley seed. Each grain of barley is a standard weight regardless of the size of the ear. A troy ounce weighs 480 grains and twelve *troy ounces* equals one pound.

In 1791, Alexander Hamilton, Secretary of the Treasury of the new United States of America, established a system of money. He favored silver coins over gold even though gold was generally desired. "One could bring fifteen ounces of silver to the U.S. Mint (in Philadelphia) and receive one ounce of gold in return." [2]

The citizens of the new country hoarded their gold or stored it in banks while silver coins circulated freely. The result was the establishment of the "silver standard." Paper money existed but coins were preferred.

Gold strikes influenced the purchasing power of gold, deflating the price nation-wide: Georgia (1828), California (1848), Australia (1852) and Colorado (1850 and 1858). Major strikes at Colorado's Central City and Idaho Springs in 1859 also affected the gold market.

Silver discoveries at Leadville, Colorado (1860-1869) brought a surplus of silver to the United States Treasury and resulting chaos. A policy was mandated permitting the printing of paper money in a direct ratio to the nation's supply of gold stored at Fort Knox. Thus the nation's finances were based on the "gold standard." The new standard was not fully implemented until 1890. The Sherman Act of 1890 mandated the government purchase of specific amounts of silver each month, and when the Sherman Act was repealed, the country was devastated by the financial Panic of 1895.

A troy ounce of gold was worth $16.00 at the time of Lewis Ralston's 1850 Colorado discovery in an Arvada stream. The price did not change until 1933 when a troy ounce was valued at $32.32. In the next few years it rose to $35.24. [3]

During the years 1930-1940 the price of gold fluctuated between $35-$43. In the 1980s an ounce of gold was valued at $840. On November 1, 2011, at the United States Federal Mint, Denver, Colorado, the price of an once of gold was $1,985! [4]

The first prospectors at any gold site usually worked with a hand-held gold pan. The gold was sifted from a pan full of water, gravel and sand. Crude rockers and sluice boxes were built from the wood of ancient trees surrounding the "diggings." Sluice boxes diverted the water from the stream bed. All early efforts were implemented by one or a small group of prospectors. The enormous bonanzas came when machinery was brought in. Large "crushers" pulverized tons of rock which was heated to release the gold. Gold production suffered in the 1900s because of the enormous costs connected with the mining process.

Modern mining has added chemicals to gold production efforts. The rock is crushed and laid out in leach pads. Cyanide drips into the pad and the chemical slowly filters through the layers, separating gold from the rock. The mixture of cyanide and gold is suctioned from the bottom of the pad and the mixture is then heated to 2,200 degrees. As the mixture cools, gold is separated from the sludge. The process recovers 62% of the gold from the ore-bearing rock. The cost of gold production is about $220 per troy ounce. [5]

In every gold rush very few seekers, neither past nor present, did or will become wealthy men. Those involved in the production process find expenses prohibitive unless the price of gold is high. The overhead margin today is encouraging and some companies are making a profit by working the "tailings," the name for rock piles left by earlier gold mining efforts. Individuals still walk the hills and dip their gold pans in the rushing rivers. There are still those who dream golden dreams.

CHAPTER II

ᘯ CHEROKEE LEGACY ᘰ
1541-1907

1541

Cherokee clans existed along the Atlantic shoreline on land that would be named Georgia and South Carolina long before the advent of European explorers. Gradually, due to fierce battles with other tribes, the Cherokee moved into the highlands. Here they were contacted on May 25, 1540 by explorers led by Spanish Captain Hernando de Soto. This encounter did not become part of Cherokee oral history, but de Soto's four scribes wrote about the interesting natives, and these records can be found in Spanish archives. These secretaries recorded a peaceful people who lived in well-constructed log houses that were roofed with bark and coated inside and out with white clay. [1]

The Cherokee Nation inhabited 40,000 square miles reaching to the slopes of the Appalachian Mountains including parts of western North Carolina, western South Carolina, southwest Virginia, eastern Tennessee, northeast Alabama and northern Georgia. The Nation was surrounded by more warlike units such as the Creeks and Chickamauga tribes. The sociable Cherokee, by 1673, had chosen a common gathering place and created a structure at a spot named Chota on the Tennessee River near present day Calhoun, Georgia. Here at Chota, a palisade of upended tree trunks, more than twelve feet high, was erected and historian Samuel Williams tells of 150 canoes tied up at the riverbank. [2]

1730

The Cherokee welcomed the first European traders. These men came from Scotland, Ireland, England and France. British and French traders often clashed as they tried to garner loyalty from the Indian market. Some traders bound themselves to tribal interests through marriage. John Adair of Ireland, married Gehoga of the Deer Clan. George Lowrey of England married Tah-nie of the Holly Clan and Englishman Edward Graves married Lahtotauyie and presented her with a spinning wheel shipped from England.

Lewis Ralston, years later, would marry Elizabeth, the great-granddaughter of Benge, chief of the Ani-Kawi or Deer Clan, Ga-ho-ga). (3)

English and French disputes over land use and trading practices led to the French and Indian War (1754-1763). Cherokee warriors fought on the side of England, while men from other tribes joined the soldiers from France. During this war, a young Captain George Washington met with Cherokee leaders and befriended Nathaniel Gist, the father of Sequoya. Gist later became Washington's representative to the Cherokee councils. (4)

1776 -1783

By the time of the Revolutionary War, many Cherokee leaders were angered by the European colonists' insatiable quest for land. Some tribal warriors fought on the side of the Americans against the British. Other Cherokee soldiers joined the English "redcoats." This group wanted the white settlers to lose the war, to lose their land usurped from the Nation, and thus defeated, sail back to England. The Cherokee had little or no under-standing of the fact that winning the War of Independence in 1783 spelled the end of their peaceful, orderly world.

1776 -1808

Chief Doublehead was the leader of the Chickamaugan tribe. He was an arrogant and bloodthirsty warrior and because of his prominence he also became a spokesman for the peaceful Cherokee. His goal was the extermination of all colonists. His trusted companion was his grandnephew, red-haired, mixed-blood Bench. Bench later became a chief, and one of his descendants was Chief John Benge. (5)

Doublehead refused to recognize the government of the new state of Georgia which had joined the Union on January 2, 1788, the fourth of the original colonies to become a state. Georgia in turn refused to recognize members of the Cherokee Nation as citizens of the United States. Georgia's leaders also subscribed to the doctrine of "state's rights," chiefly, the right to own slaves. Doublehead was part of an Indian delegation that was taken to meet President Washington in 1794. This delegation signed a treaty which paid the Cherokee $1,500 for the land that had been settled by colonists. The tribes were not pacified and in 1808, Doublehead and 1,130 of his followers moved to lands that became the state of Arkansas in 1830. (6)

Before the removal of the Chicamaugans, Doublehead was persuaded to give away two parcels of Cherokee land. He was bribed by Return Jonathan Meigs and James Vann, the latter the owner of a mill and ferry on the Consaga River. Vann was eager to own the property because a federal road was being constructed that would connect with his property. The disposal of land within the Cherokee Nation, land owned in common by all, was treason according to Cherokee leaders, and in 1808, three men, Major Ridge, Alex Saunders and John Rogers murdered Doublehead.

Although the Cherokee were many separate clans, each headed by a "chief," Bench emerged as a strong unifying leader. He founded a Legislative Council and this body developed rules to which everyone in the Cherokee Nation could subscribe. The Council actions were not in line with the dictates of Georgia or the federal government. In 1825 they established a capital, New Echota, at the confluence of the Coosawattee and Conasauga rivers. At about this time, Cherokee individuals, who through the ages had been known by a single name, began to use two names and Bench became Chief John Bench.

The existence of a separate nation within the state of Georgia became increasingly divisive. Sequoya had invented a Cherokee alphabet and was teaching adults to read. At New Echota, the Nation printed a newspaper in 1828, the *Cherokee Advocate,* which helped encourage a spirit of unity for the Nation, binding them together to resist the dictates of state and federal government.

1828 ~ GOLD

The discovery of gold in 1828 (see chapter III) triggered the end of the power of the Cherokee Nation. One year later, President Andrew Jackson, in his address to the United States Congress on December 8, 1829, called for the removal of the Cherokees to the West, to "lands beyond the Mississippi." He promised funds (which proved to be insufficient) to make the removal a reality.

In the early years of the 1800s, John Ross had assumed the role of leader of the Cherokee Nation. He had attended schools in the North and was an experienced attorney. He tried to convince the federal government that the removal was wrong. If the United States would not accept the Cherokees as citizens, he argued, then they should be given the common courtesy of negotiation that was extended to foreign governments. He succeeded in presenting his case to the U. S. Supreme Court. This high authority said both Georgia and the federal government were wrong, but this judgement was ignored by President Jackson.

Chief Ross returned to Georgia in April of 1833. He found that his property had been taken over by several winners of the 1830 Georgia Land Lottery. His ailing wife Quatie and his children were locked in two rooms of his beautiful home. He sent them to a friend in Tennessee.

John Ross called for a meeting of his supporters in July of 1835. At that time, his position as Principal Chief of the Nation was reaffirmed and the future of the Georgia Cherokee was placed in his hands. The attempt of John Ridge to assume the role of Principal Chief was denied.

Seven years after the gold discovery, U. S. officials met a group of Cherokees at New Echota on December 29, 1835. Some 100 members of the group signed a new treaty, but these individuals were not representatives of the Nation. The signers agreed to the sale of all Cherokee land for the sum of $5 million dollars. This group, later known as the Treaty Party, left almost immediately for lands in the West (Oklahoma). The organizers of the treaty signing, John Ridge, his father Major Ridge and Elias Boudinout were savagely killed by furious Cherokee loyalists in 1839. The fourth member of the group, Stand Watie, fled the area.

With the signing of the New Echota Treaty, the fate of the Eastern Cherokee had been sealed. Chief Ross had to assume the task of protecting his people through the process of the forced removal.

1838 -1839

The sad account of the forced exodus is known by the Cherokee as "the trail of those who cried," and to history as the Trail of Tears. Beginning with the summer of 1838, Cherokee families were driven from their homes by the soldiers of U. S. General Winfield Scott. From the stockades they were forced into badly provisioned wagons. Some rode their own horses and others walked, all moving West. The move was so haphazard, with so few Cherokee actually reaching Indian Territory, that the government appointed Chief John Ross to direct matters. He organized his people into companies each headed by their own officers. The first detachment left in October of 1838, headed by Chief John Benge and George Lowrey. Quatie, the wife of John Ross, died on the journey and was buried near present-day Little Rock, Arkansas. By the spring of 1839, according to historian Grant Foreman, "18,000 Cherokees had been removed from Georgia and of these 4,000 had perished along the way." (7)

1839

The new Indian Nation, later the state of Oklahoma, was home to Cherokee, Creek, Potawatomi, Choctaw and Chickasaw tribes. Each group gathered in a separate village and tried to rebuild their way of life. The Cherokee were divided into the Treaty Party, the Old Settlers who were first to remove, and the "Eastern tribes." The latter were the survivors of the Trail of Tears. After much turmoil and compromise, John Ross was elected Principal Chief of the Cherokee Nation West and the capital was established at Tahlequah. Chief Ross tried to keep his people free of participation in the Civil War, but there was division even in this new nation over the war to end slavery. Following the end of the war, a period of relative peace prevailed for the next decade.

1848 ~ GOLD

Gold was discovered in what would become the state of California on January 24, 1848, twenty years after the Georgia strike. James Marshall, helping to construct a sawmill for Captain John Augustus Sutter, found small gold nuggets "half the size and shape of a pea," in the under-construction mill-wheel channel. (7)

Peter and Jenny Wimmer were part of the sawmill crew and Jenny wrote home to Lumpkin County, Georgia, telling of the discovery. The news quickly spread throughout the thirty states of the Union and triggered a stampede of gold seekers called "the 49ers." Many of the wagon trains stopped at Tahlequah or other trading centers to rest, to repair wagon wheels or to replenish their stocks, bringing prosperity to some of the enterprising merchants of the Indian Nation.

The California gold strike, the second major gold find in the United States, led to the beginning of the third major gold rush which occurred in a mountain stream still rushing today through Arvada, Colorado. (see Chapter IV)

1893

The United States Congress established the Dawes Commission on March 3, 1893. Federal agents endeavored to register every Cherokee individual and/or household. Reluctant Cherokees at first did not cooperate until they found that this would be a way to establish legal ownership of land. More than 300,000 individuals applied for registration and of this number, 100,000 could support their claim for Cherokee ancestry. In 1902, each enrolled member was entitled to 110 acres of land. Cherokee leaders applied for territorial status for Indian Territory, which was never adopted because the federal government demanded the end of the Cherokee Nation's own rules and laws. The Cherokees finally dissolved their own historical government in 1906. The various tribes living in the Indian Nation banded together and Cherokee lands were included in the petition for statehood.

Oklahoma became the 46th state by proclamation of President Theodore Roosevelt on November 16, 1907. More than 69 long years had elapsed since the beginning of the Cherokee removal from Georgia in 1838.

1992

More than 160 years after the beginnings of the Cherokee tragedy, Georgia legislators apologized for the imprisonment of two white missionaries who were sent to jail in 1832 for speaking out about the usurpation of Cherokee lands. The following clipping from a Colorado newspaper, the *Denver Post,* closes the Removal story.

Members of the Cherokee Nation today are proud of their heritage and are a strong presence in Oklahoma and the nation. They are also proud and steadfast citizens of the United States.

Georgia admits mistake in seizing Cherokee land

By The Associated Press

ATLANTA — More than 160 years after Georgia officials ignored a direct order from the U.S. Supreme Court to stop actions leading up to the infamous Trail of Tears, the state is admitting it made a mistake.

Officials on Wednesday will formally pardon two missionaries jailed when they fought the state's seizure of Cherokee Indian land.

The pardon says it "acts to remove a stain on the history of criminal justice in Georgia" and acknowledges that the state usurped Cherokee sovereignty and ignored the Supreme Court.

A legislator and Cherokee descendant called the pardon a sign that Georgia finally realizes the scope of its mistreatment of the Cherokee.

"If we ever had political prisoners in this state or this nation, these two (missionaries) were the best examples," said state Rep. Bill Dover, Georgia Tribe of Eastern Cherokee chief executive.

Samuel Austin Worcester and Elihu Butler were sentenced to four years in jail in 1831 for residing in the Cherokee Nation without a license. A law was enacted to try to stop the two from protesting the state's seizure of Cherokee land.

Until 1828, the Cherokee Nation was considered a sovereign foreign country. But the next year, gold was discovered in Dahlonega, and Georgia seized much of the land.

Worcester and Butler attracted national attention to the Indians' cause. To muzzle them, the state required all white men living on Cherokee land to obtain a state license. The two refused and were convicted of "high misdemeanor."

The missionaries appealed to the U.S. Supreme Court. In 1832, Chief Justice John Marshall declared Georgia had no constitutional right to extend any state laws over the Cherokee.

But Georgia ignored the ruling, and the missionaries spent 16 months doing hard labor. They were released in time to join the Trail of Tears, when Georgia forced 17,000 Cherokees to move west.

Painting in Heritage Center, Salt Lake City, Utah

19

ᴽ᷎ LEWIS RALSTON & GEORGIA GOLD ᷎ᴽ

Lewis Ralston was born in 1804, the son of John Tate Ralston and Lettie Harris. John Tate Ralston was born in Great Britain, probably in Scotland. John and Lettie settled on a farm near what would become Pendleton, South Carolina. (1)

The small village of Pendleton was part of the more than 2,000 mile spread of territory claimed by Cherokee Indian tribes. Lewis grew to manhood with Cherokee families as neighbors and friends. At the age of 21 he traveled 100 miles, crossed the wide Savannah River, and reached what would later be Lumpkin County, Georgia, part of the vast Cherokee Nation.

According to the *Gazetteer,* published in 1837: "Sir Walter Raleigh is the reputed discoverer of the territory now called Georgia." He however did not found a colony in Georgia. The first English settlements were Jamestown, Virginia in 1607 and Plymouth, Massachusetts in 1620. These colonies kept lists of residents, but settlers like John Tate Ralston usually were not listed on historical documents. The first Georgia colony was established by James Oglethorpe in 1732-1733. He and 120 brave followers founded the town of Savannah. (2) However, before 1732 the Cherokee had built towns of their own – towns with names like Catoosa, Tugaloo and Echota. Echota was the center of tribal government, located on the Coosawattee River in an "untrodden forest." It was the Cherokee heart for more than twenty towns and villages. (3)

When Lewis Ralston in 1825 reached the area that would be named Lumpkin County, he met Benjamin Parks Jr., and they became partners as suppliers of cows and horses. Parks was the son of Benjamin Parks, Sr. who immigrated from Scotland and fought on the side of the colonists in the Revolutionary War. The family, originally from Virginia, eventually moved to Georgia. They had spent some time in North Carolina and here Benjamin Parks Jr. married Sarah Henderson. (4)

Although Benjamin Jr. and Lewis lived outside the Cherokee Nation, they had permission to graze their livestock through the cool green forest west of the Chestatee River. The river marked the boundary of the Cherokee Nation. They had a "lick-log" in a clearing – a log that was hollowed out and filled with salt to improve the health of their animals. Parks said about the Cherokee residents, "(my) family felt safe and welcome among them."

On a warm autumn day in 1828, Parks was out deer hunting. His daughter wrote his account of the day: "Crossing a little dried-up water-course, I kicked up a nice quartz piece with a sparkle to it that caught my eye. When I looked I knew it had to be but only one thing – gold!" (5)

News of this discovery spread like wildfire and repeated telling resulted in a confusion of dates and details. Parks said: "It was my birthday so I'd ought to know (October 27, 1828). And I'm telling you – the to-do that followed was going full stream (steam) before ever 1829 was rung in." Parks later said, "it seemed within a few days as if the whole world must have heard of it for men came from every state I have ever heard of . They came afoot, on horseback and in wagons, acting more like crazy men than any-thing else." (6)

Parks found gold on the land that Robert O'Bar, a Baptist minister had leased from the Cherokee Nation. O'Bar was building the Yellow Creek Baptist Church, but agreed to give Parks a forty-year lease on part of his leased property. Parks found more rocks with golden streaks "like the yellow of an egg." He and Lewis Ralston began establishing a mine and they hired other friends to help build a sluice box. O'Bar had second thoughts and tried to break his contract with Parks. Parks refused. O'Bar and other church members destroyed the work of the Ralston/Parks team.

Senator John C. Calhoun from South Carolina visited the site in 1828 soon after Parks' team began re-building their sluice. Although he had assured the Cherokees that he was their friend, he immediately bought the church lease from O'Bar, and Calhoun's armed workers drove the Parks' group away from their new gold mine. The Calhoun mine became one of the major producers of the gold region sending 24,000 pennyweights of gold ($23,000) to the U. S. Mint in Philadelphia in the first month of operation. (7)

A settlement grew almost overnight on the high ridge between the Chestatee and Etowah rivers, land that had been, by treaty, awarded to the Cherokee Nation. William Dean was the first settler in 1829, followed only days later by Nathaniel Nuckolls. Nuckolls built a large log house to serve as a tavern and living space for his family. As more settlers arrived, the town was called Dean's or Nuckollsville. Senator Calhoun became Vice-President of the 24 states of the Union under President Andrew Jackson in 1829. He visited his mine and decided the new gold mining settlement needed a name. He announced that the new town should be named *Auraria,* a Latin word meaning gold.

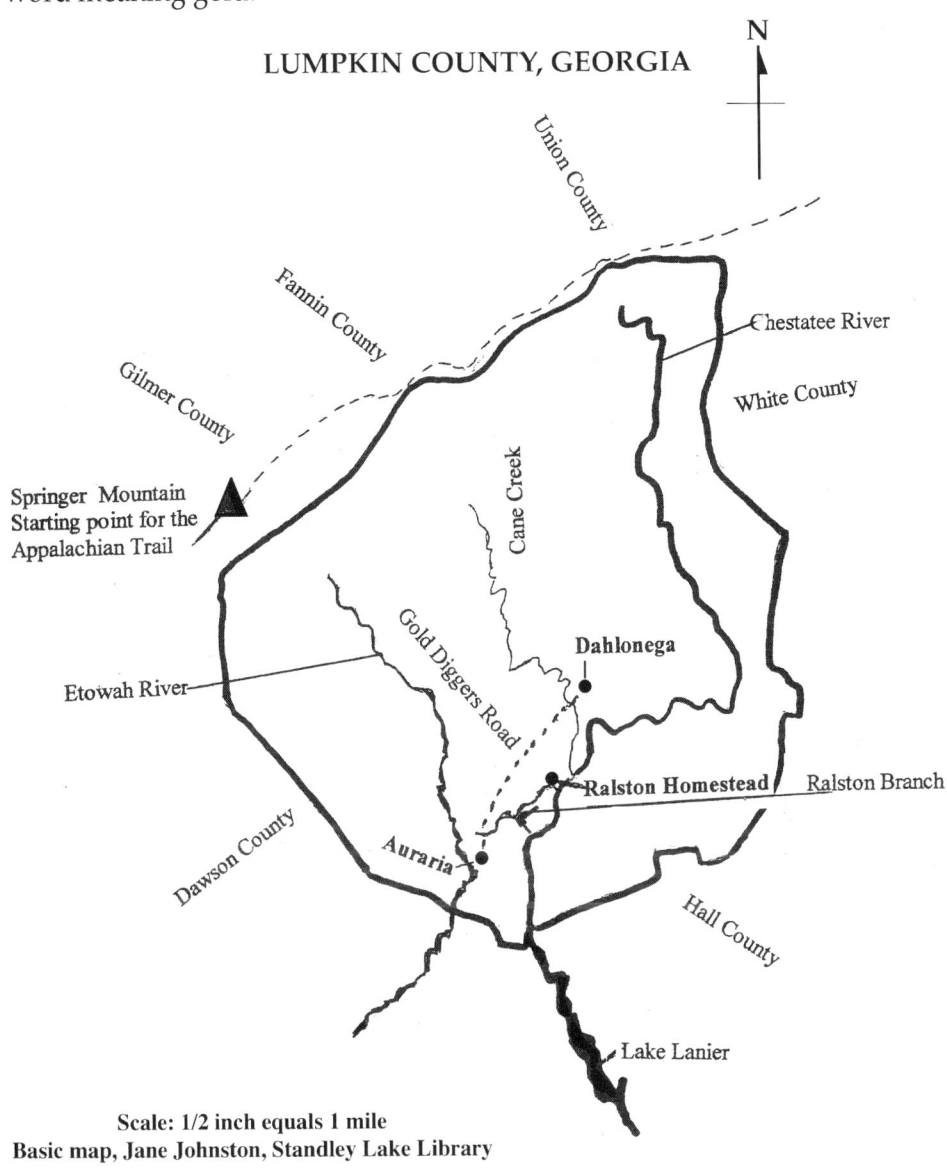

LUMPKIN COUNTY, GEORGIA

Scale: 1/2 inch equals 1 mile
Basic map, Jane Johnston, Standley Lake Library

Research Lois Lindstrom

Leaders of the Cherokee Nation were powerless to stem the flood of gold seekers. Historian Anne D. Amerson said: "by June of 1830 there were 4,000 miners working the banks of Yahoola Creek." Three years later Auraria was a village of 100 log cabins, most certainly inadequate housing for the prospectors crowding the area. (8)

Cherokees were not permitted to live within a two-mile radius of Auraria and could not remain in the town overnight. The unfortunates, who had established a home on the site were evicted by the ruthless gold seekers. The Cherokee had no voice in town government. There were many saloons and Auraria became a rough and ready boomtown.

In 1833, approximately five miles north of Auraria, a new town was built also on Cherokee land. It was named *Dahlonega* after an Indian name Talonega which means gold or gold mine. Dahlonega leaders succeeded in receiving the designation of county seat for Lumpkin County, and in 1834 a courthouse was built on the site of the Ralston/Parks "lick-log." The trail between the two towns was called Gold Diggers Road.

Lewis Ralston met the great-granddaughter of Cherokee Chief John Benge, Chief of the Ani-Kawi (Deer Clan). Benge, an influential Cherokee leader, was a descendant of Chief Bench, the grandnephew of fierce Chief Doublehead. We have no description of Elizabeth Duncan Kell, but Parks was quoted as praising the beauty of Cherokee women. Lewis, and Elizabeth were married in 1825 and he became known as an Indian countryman, that is a white man married to a Cherokee woman. They settled on a farm near the confluence of a small stream and the mighty Chestateee River and the stream was named after him – Ralston Branch. Lewis built a sturdy "double log house (1 ½ stories of hewed logs) a common house, another small cabin, two stables a smokehouse, a chicken house and two corn cribs." (9)

Ralston built a ferry landing and his slaves managed this raft crossing of the Chestatee River. The farm included a thriving peach orchard. Lewis was recognized as one of the leading planters in Lumpkin County. He probably raised corn for sale and to fatten his cattle. Soon after Parks' discovery, Lewis established his own gold mine.

Ralston had purchased land on Yahoola Creek before the discovery of gold. He also owned land on Hightower River, as well as properties on the Etowah River near Bread Town and land near Tensawatee Town.

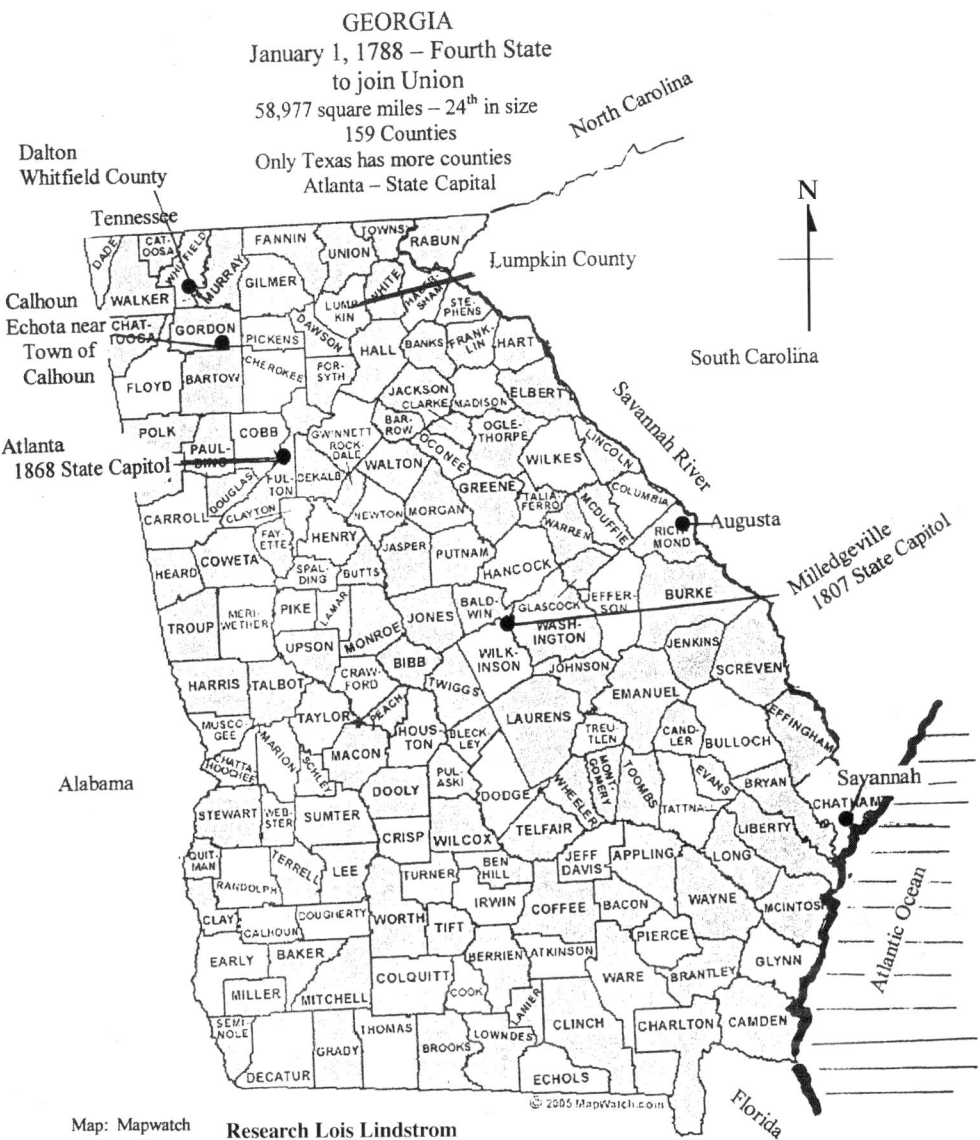

GEORGIA

January 1, 1788 – Fourth State
to join Union
58,977 square miles – 24th in size
159 Counties
Only Texas has more counties
Atlanta – State Capital

Dalton
Whitfield County

Tennessee

North Carolina

Lumpkin County

N

Calhoun
Echota near
Town of
Calhoun

South Carolina

Atlanta
1868 State Capitol

Savannah River

Augusta

Milledgeville
1807 State Capitol

Alabama

Savannah

Atlantic Ocean

Florida

© 2005 MapWatch.com

Map: Mapwatch   **Research Lois Lindstrom**

Some Cherokee families, either willingly or by coercion, were moved from Georgia as early as 1836. Lewis filed a claim for prospector-caused damages to his property. His total claim was almost twice the salary of the President of the United States who was paid $3,000 a year. Ralston claimed $5,894. (10) There is no record that his claim was honored.

Andrew Jackson, who became President of the United States in 1829, was no friend of the Cherokee Nation. The state of Georgia could not control the actions of greedy prospectors, and the state appealed to the U. S. Congress for help. Jackson spoke to the federal legislature and insisted that removal of the Cherokee, Creeks and other native tribes was the only solution. Jackson advocated re-settlement to some location in the far West. The Georgia Cherokees were stunned to think they would be moved to unknown lands. Their land ownership had been affirmed over and over again in treaties signed by the United States.

As early as 1829, only months after Parks' gold discovery, the federal government developed a plan that would divide the Cherokee Nation in Georgia into ten large counties. One of these was named Cherokee County. In 1830 Cherokee County itself was divided into smaller counties. Lumpkin County was created, the area of the richest gold mines. The county was named for Wilson Lumpkin, who had been elected Governor of Georgia in 1829.

On June 1, 1830 the Georgia legislature passed laws suppressing all council meetings of the Cherokee Nation and "all (Cherokee) functions, laws and customs were made null and void."(11) The Georgia state militia were given a free hand to punish any disturbance or outcry.

The Cherokees were as civilized, perhaps even more so, than the gold seekers who swarmed into Lumpkin County. Many Cherokee could read and write both English and their own language. In the 1800s Sequoya had developed a Cherokee alphabet and the *Cherokee Phoenix* newspaper, with text in both English and Cherokee, was published weekly from New Echota. The peaceful tribes included businessmen, farmers, lawyers and artisans. Their only crime was occupancy of gold-bearing land, and for that, their punishment was just beginning.

In 1832, the Georgia legislature began dividing counties into lots of 160 acres each. In "gold" counties like Lumpkin County, the lots were limited to forty acres. The Georgia Land Lottery was then established allowing only white men to enter. Each signer was asked to certify that he had resided in Georgia for four years, but many men falsified their years of residency. Choice lots were given to veterans of the Revolutionary War or as a reward to individuals for past political favors. Cherokee families were evicted and lottery winners moved in.

On December 29. 1835 the Treaty of New Echota was signed by 100 Cherokee men who were not elected representatives of the Cherokee Nation. This treaty illegally sold the lands of the Eastern Cherokee Nation for the sum of $5 million dollars. Most members of the Nation did not receive any benefit from the dollars paid. Lewis Ralston and Joseph Lynch, listed as "attorneys" tried to break the terms of the treaty saying that the signers were not legal and their signing was invalid. This case probably came to a judge in Milledgeville, the town chosen as Georgia state capital in 1807. The capital was "in the heart of cotton country," several days journey from Dahlonega. Ralston and Lynch were unsuccessful. Signers of the Treaty of New Echota, with their fortune, left almost immediately for the West and settled in what would become Oklahoma. This group was known as the Treaty Party.

Chief John Ross had assumed a position of leadership for his people. He led a delegation to Washington D. C. trying to stem the tide of the forced immigration. He argued with federal leaders saying that if the Cherokee could not be treated as American citizens, that they then should have the status of a foreign country and be allowed to co-exist with the United States just, as did the citizens of Canada. Ross, a certified attorney, paid all legal expenses himself and managed the case through lesser courts all the way to the federal Supreme Court. That body stated that Georgia did not have the authority to overrule federal laws, and that "forced removal" was contrary to laws of the United States. President Jackson simply ignored the verdict.

Some evicted Cherokees on their own, moved to the far West. They, like the Treaty Party, settled on fertile land near lakes or rivers, acquiring some of the best land in what would become Oklahoma. These immigrants were the nucleus of the Old Settlers Party. In 1817, more than 2,000 Cherokees moved to Tennessee. Other small groups left Georgia in 1818 and 1819.

Lewis Ralston was determined to save his gold mine and his farm. On October 7, 1830 he signed the Oath of Allegiance to the United States of

America. He thought this action would protect his properties and save his gold mine located on Ralston Branch, Lot 837. In the 1832 Land Lottery, his gold mine was awarded to Henry Slaughter. Lewis furiously assembled his family and slaves, and removed "6 thousand pennyweights of gold." Slaughter took the case to court and Lewis Ralston was fined $6,000. (12)

> A pennyweight is equal to 1/20th of a troy ounce.
> In 1830 a troy ounce of gold was worth $19.
> Thus, a pennyweight was worth less than $1.
> Ralston's fine was more than the amount of
> gold he mined from Lot 837.

It is difficult today to locate Lewis Ralston's gold mine. Librarians at the Dahlonega Gold Museum found the following information: "The property was worked first as a placer along Ralston's branch between 1840 and 1845 by (owner) Elisha Castlebury." There is no record of Ralston's or Slaughter's ownership, destroyed no doubt when all Cherokee records were obliterated. Castlebury owned lots 726, 728 and 731 and left information that the gold content of the Ralston's Branch gravels was "very rich." (13)

Lewis Ralston owned lots 808, 871 and 880 and these lots were transferred to John Galloway in 1842. A man named Stephen Crane, in his will, left "three eighths of (lot) 83-12-1" to Lewis Ralston. (14)

Dahlonega succeeded in securing a clear title to land and the post office was awarded to the new town. Auraria could not contest the decision because land they favored could not furnish a clear title. Dahlonega built a large log post office building. It was erected in the clearing where Parks and Ralston had placed their lick-log. A strong brick post office building replaced the log cabin three years later. Gold mining was very successful and $1,763,900 had been sent to the U.S. Mint in Philadelphia by 1837. The Dahlonega Mint was built the same year. (15)

Chief John Ross returned from Washington in 1933, discouraged and resigned to the perilous removal facing his people. He found several Land Lottery winners living in his barn and his beautiful home. His wife, Quatie, and his children, were locked into two rooms of the house. He comforted them and sent them with an escort to the home of a friend in Tennessee. Shortly, the soldiers of the United States Army, under direction of General Winfield Scott, were herding Cherokee families into stockades where there was almost no provision of food, medicine or blankets. The money the federal government had advanced to finance the forced move, inadequate to start with, had been ill spent. Many Cherokees, riding ponies or in poorly constructed wagons, were being sent West with almost nothing to sustain life.

The federal government recalled General Scott and his army and asked Chief John Ross to supervise the move. The weather was unusually cold and the Cherokee themselves were in poor condition. The Trail of Tears had begun.

Lewis and Elizabeth were among the residents of Dahlonega who had managed to evade the dictates of the federal government. They still had their home and farm but Lumpkin County was steeped in sorrow. In 1848 the Ralstons were among the first to hear of a letter sent from the shores of the Pacific Ocean.

Elizabeth Jane Cloud, always called "Jenny," had moved with her family to Auraria, Georgia in 1838 when she was 16 years old. She married a man named Peter Wimmer and they traveled the Oregon Trail west. They both found employment with Captain John Augustus Sutter who was building his empire in Mexican Territory. Jenny's letter to her family in Auraria told of a gold find made by James Marshall on January 24, 1848. She said that some nuggets were "as large as a pea" and that prospectors were rushing to the area and building the town of Coloma. (16)

The men of Lumpkin County realized they had a chance to be at the beginning of another gold rush. A meeting was held on the courthouse steps in Dahlonega and plans were made for prospecting parties to start west. Greeneberry Russell, known as "Green" became the leader for one group. The Russell family lived at Gainsville in Hall County, Georgia – about 20 miles south of Auraria. Green was an excellent leader. He was familiar with the Oregon Trail (opened in 1811) and had visited friends in the new Cherokee Nation (Oklahoma). Lewis Ralston listened to the eager prospectors. Green and his party left Lumpkin County in 1849, but Lewis was held back by his family responsibilities. A book written by one of Green's descendants does not have a date for the beginning of Green's prospecting journey. She, the author, uses facts from the John Lowery Brown diary to tell of the trip. The diary is documentation of the McNair prospecting party and was not written until 1850.

In April of 1850, more than a year after Jenny's letter arrived, Ralston decided once again to be a prospector. He joined the wagon train led by Captain Clement Van McNair who promised to lead the group to California (statehood achieved in September 1850).

By 1850, Lewis and Elizabeth were the parents of 13 children. He had suffered loss of property and wealth in the confiscation of Cherokee Nation lands. He had signed both the Siler and Henderson lists, and because he was a white man, his family was able to retain their farm on Ralston Branch. His family agreed that he should try once again to make his fortune, and his journey is detailed in Chapter V.

Ralston again responded to the siren song of gold.

Remnants of history live on in old mining and
ghost towns throughout the state.

CHAPTER IV

⋙ LAND OF GOLD & GRIEF ⋘

There are no photographs of Lewis or Elizabeth Ralston and there are no written descriptions of them. Removal lists, court actions and census records are the only documentation for the Ralstons.

Lewis arrived in the Cherokee Nation at the age of 21 in 1825. He and Benjamin Parks became partners and he married Elizabeth Kell in 1826. They established a farm at the confluence of the Chestatee River and a small stream that soon would be named for Lewis – Ralston Branch. Nearby was a village named Old Chestatee Town. The farm was east of the river.

Elizabeth was the daughter of Emily Duncan and Alexander Kell. Alexander was a white man whose ancestors came from Scotland. The mother of Emily Duncan Kell was Dorcas, a Cherokee princess and the daughter of Chief Benge (later Chief John Benge). Dorcas first married Captain John Lightfoot and after his death she married Charles G. Duncan. Their daughter was Emily Duncan. Emily's daughter, Elizabeth Kell Ralston, thus was the great-granddaughter of Chief Benge, the leader of the Ani-Kawi, known as the Deer Clan (Ga-ho-ga). [1]

Lewis was accepted by Elizabeth's family and participated in the orderly routines established by the Cherokee constitution and rules. He was known as an Indian Countryman. Their first child, Rebecca, was born in 1827. Lewis, due to his partnership with Parks, was one of the first to learn of the 1828 discovery of gold on land leased from the Cherokee Nation. He no doubt helped, with others, to build a wooden sluice box to capture gold deposits from the waters of the Yahoola River. Parks would have asked Ralston to help when Calhoun's armed guards drove the miners away from his claim. Ralston and his Cherokee friends were furious when the gold camp town of Auraria sprouted on Cherokee land, and when that town would only allow them entry for a few hours each day. Six miles north a second town, Dahlonega, was established and these town leaders succeeded in securing the designation of county seat. The Cherokees were not banned from Dahlonega.

During the early gold rush years, the intruding prospectors, who disregarded the ownership of the tribes, promoted a greater unity among the clans of the Cherokee Nation. Tribal leaders sent appeals to the State of Georgia asking for protection from the gold seeking invaders. Their requests were ignored. During this time, according to some Ralston family history writers, a son of Benjamin Parks married Lewis Ralston's sister Clarissa. However, Benjamin Parks Sr. did not name any of his sons Benjamin, and his grandson, Benjamin Parks was not born until 1850. (2)

Soon after Parks' discovery, Lewis Ralston panned for gold himself on Ralston Branch, and he developed a placer gold mining operation. Both men were amazed when Lumpkin County was created and the town of Dahlonega in 1833 built a log structure to serve as the courthouse. The courthouse steps were built where they had placed their lick-log during their livestock days and the town itself was nicknamed "Licklog" until the name Dahlonega became official. In 1836 the log cabin courthouse was replaced by a handsome brick building that today houses the Dahlonega Gold Museum.

Many of Georgia's famous locations are in the area claimed by the Cherokee Nation. A line drawn north to south following the eastern boundary of the Nation, divides the state into two almost equal parts – the western part known as Cherokee land. The tallest mountain in the state at 4,784 feet is located in the north and is part of the beautiful misty Blue Ridge Mountains. The starting point for the Appalachian Trail is in Georgia. Northern Georgia is home to green forests of pine and white oak. The red clay of the forest floor is nourished by many small sparkling streams. Dogwood trees and bushes of the Cherokee rose bloom in profusion. Both Georgia and the federal government coveted the Cherokee lands.

Late in 1830, Lewis Ralston realized that Cherokee removal was inevitable and he acted to protect his own properties. On October 7, 1830 he signed the Oath of Allegiance to Georgia and the United States. Two months later the Georgia General Assembly met in the capital city, Milledgeville. The lawmakers proclaimed that the State of Georgia now owned "all the territory within the limits of Georgia and now in the occupancy of the Cherokee tribe of Indians." (3)

By 1833 the Land Lottery had been fully implemented. Lewis engaged in court actions to prove that he as a white man was entitled to own Georgia property. He served as his own attorney.

Lewis had been assured that he could remain in Georgia and that he could retain ownership of some of his property. He was shocked to find his name on the 1835 Henderson Roll of those who would be forced to move to Indian Territory in the far west (Oklahoma). David Henderson compiled his list from the 1835 census:

> "Lumpkin County has extensive goldmines, iron ore, etc. Has extensive water power and about one-fifth of the land is tillable.
>
> Lumpkin County – Etower (Etowah) River Lewis Ralston Five quarterblood Indians, one white marriage. They owned a ferry boat. One farmer, 2 readers of English. 1 weaver, 1 spinster. 5 descendants of reservees." [4]

To translate Henderson's report: Lewis would be the farmer, perhaps he and Elizabeth were the readers of English, and the "5 descendants" are the children – Rebecca, John, Alexander, Frances and Emily, ranging in age from Rebecca, age 8 to baby Emily, age 2. It is possible that Elizabeth was the weaver but the person listed as "spinster" is a mystery. A spinster in the 1800s was a single woman who had passed the conventional age of marriage. This might be the age of 14, or 16. Rebecca, the oldest daughter was 8 years old.

More than 45 years earlier, a Lewis Ralston is listed on the 1790 census of Pendleton County, South Carolina (where Lewis Ralston of Lumpkin County was born). Was the Ralston of South Carolina listed 45 years earlier a relative? The Ralston of South Carolina owned six slaves, and perhaps Lewis of Georgia inherited slaves from him. There is no answer to the question.

Henderson's report of 1835 was a great worry to Lewis and Elizabeth because it indicated that they would be part of the forced exodus. They had read a letter written in 1830 by R. Montgomery to the Cherokee Agency. The letter said that the Lumpkin County Ralstons were on the list of those individuals who had permission to remain in Georgia. Perhaps Lewis contacted Henderson to inform him that his family would not move.

Lewis was involved in court actions trying to defend his title to his various properties and also trying to protect the rights of all Cherokees who would be forced to migrate according to the Treaty of New Echota. He made many trips to the Lumpkin County Court House in Dahlonega. In 1832 Lumpkin County had been named for the Governor of Georgia, Wilson Lumpkin. Lumpkin had served in both houses of the United States Congress and had been instrumental in passing the laws that would force the Cherokee from their native land. Lumpkin was not a pleasing word in Cherokee ears.

Through these tension-filled days many Cherokee found comfort in their churches. Protestant missionaries had been working for years in the Nation. The Cherokee councils had allowed northern congregations to establish local churches in their Nation only if they first opened schools for the children. One of the best-known schools was that established by Presbyterian Pastor Gideon Blackburn. There are no records to show Ralston church membership. They might have attended the Yellow Creek Baptist church built near Benjamin Parks' first gold find. They may have attended the Antioch Baptist Church of Auraria. The Auraria church had been founded by Mrs. Agnes Paschal, and one of the Ralston daughters was named after her. (5)

In 1835, the state government began the process of surveying and evaluating the properties of the Cherokee Nation. The following record was found in the Book of Valuation, December 20, 1835:

> "Lot number 102, November 1, appraised for Lewis Ralston, a white man and his wife Elizabeth Kell at the Chestatee Old Town on the Chestatee River where Ralston now resides in Lumpkin County–
>
>> 1 large double hued log house 1 and 1/2 stories
>> 1 large double hued 1 story common
>> 1 cabin 1 smoke house
>> 1 stable 1 stable 1 crib
>> 1 crib 34 peach trees
>> 92 acres upland at $800
>> 300 rad ditching for very deep ----- (?)
>> 1 poultry house
>
> Total $2,037.00"

Additional properties were listed on the Yahoola River, on the Hightower (Etowah) River, the Amicalola River and in Terrwatter town. (6)

The Trail of Tears

By 1838 some of the Cherokees of Georgia had resigned themselves to the loss of their ancient home and had on their own or in small groups moved westward. There were many who refused to leave their beloved homes. The United States government sent General Winfield Scott and his soldiers to the assistance of the State of Georgia. General Scott proceeded to occupy the ancient historic Cherokee Council House at New Echota, and converted the building to army barracks for his troops.

Armed with rifles, bayonets affixed, the soldiers stormed into the homes of terrified Cherokee families. These unfortunates were prodded out of their dwellings, carrying only what few treasures they could push into homspun cloth bags. They were not permitted to bring any extra clothing or blankets. The unarmed people were placed in poorly constructed stockades, fed unclean food and provided with little water. Many sickened and died. After a few months of chaos Cherokee Chief John Ross was asked to take charge of the removal.

Chief Ross was in Washington D. C. pleading with Congress and the Supreme Court to halt the move. Convinced that his cause was lost he agreed to try to bring an element of human care to the forced migration. He returned to Georgia in 1838 to find his home occupied by lottery winners, and after sending his family to Tennessee, he found lodging in a one-room log cabin. Ross made lists and ordered wagons to be built. He divided his people into 13 parties of 1,000 each. The first detachment left for Indian Territory (Oklahoma) in October of 1838. He asked Chief John Benge (Elizabeth Ralston's great-grandfather) and George Lowrey to help him and they served as officers for the first group. This first party included many of Elizabeth's relatives. Ross also asked that the weakest and sickest go with this first party because he knew they would die if forced to live any longer in the stockades. Of the 1,103 in the detachment, only 489 lived to complete the journey. Cherokee Historian Grant Foreman says a total of 18,000 people started the trip to the west and of that number 4,000 perished along the way. (7)

It is hard to accept that the people of the then 26 states of the Union did not protest this brutal treatment of a peace-loving people. The townsmen of Marietta, Ohio were the only group to openly denounce the actions of the government. However, the turmoil over slavery consumed the hearts and minds of the Nation, resulting finally, in 1861, with the beginning of the Civil War. By 1840 the states and their representatives were beginning to line up on either side of an invisible wall that divided the United States into positions of either "free" or "slave."

The question of slavery had been a problem since the inception of the nation in 1776. On January 1, 1808 the United States Congress passed a law prohibiting any further importation of slaves. However, in the southern states the large plantation owners continued to use, buy and sell their slaves. James Pierce Butler, second on the list of Georgia slave owners, who controlled the lives of more than 1,000 slaves on various Georgia plantations, actively promoted the continuation of slavery. Butler and other slave owners had powerful friends in Congress. In 1820, the Missouri Compromise was instituted which tried to balance the slave question when new states joined the Union. Maine entered the nation as "free" and Missouri as "slave." A line was drawn across the map of the country, which continued the southern border of Missouri westward, and only south of this line could a new state declare itself "slave." The truce did not last.

Abolitionists became increasingly vocal. The first Anti-Slavery Convention was held in New York City on May 1, 1837. The slavery question dominated government and political debates. Little attention was given to the removal of Cherokees and other native tribes to the arid lands of the West.

Elizabeth Ralston wept at the loss of her family. She and Lewis sorrowed over the empty houses and desolate streets of Lumpkin County. They soon realized that their peaceful Cherokee existence would undergo radical change as railroads were constructed into the now vacant Cherokee lands. The first Georgia railroad was constructed in 1832 which served traffic from Augusta to Athens. In 1837, even before Removal, a line was proposed to a small village called Peachtree in the former Cherokee Nation. Soon two lines were built to the village, re-named Terminus. It was re-named Atlanta in 1845 and in 1867 became the capital city. The dome of the State Capitol was gilded with Lumpkin County gold. (8)

The Ralston Gold Mine

Lewis Ralston's mine on Ralston Branch has kept its name for almost 180 years. It has been owned by many including a large consolidated firm. The Ralston name has persisted although Lewis himself lost the gold.

In 1830 all Cherokee rights were terminated in Georgia. All records were destroyed. This included Lewis Ralston's ownership of land, even though he was not a Cherokee. There is no existing documentation for Lewis Ralston's ownership of a gold mine on Ralston Branch.

The gold mine originally consisted of lots 726, 728 and 731 and it adjoined the claim of the Barlow Mine. Modern records for the mine begin with the year 1840 (2 years after the Removal) and state: "the mine was worked first as a placer along Ralston's Branch . . . by (owner) Mr. Elisha Castlebury." There is no mention of Lewis Ralston. Mr. Castlebury's son is quoted as saying that "it was a very rich mine." Lewis Ralston's contribution was simply erased from history.

The Ralston Mine is one of the nine properties included in the 4,600 acres of the Dahlonega Consolidated Gold Mining Company which folded nine small mines into one operation in 1898. The company was lauded in 1899 for "employing every industrious laborer with or without a team." In 2008 the mine was open for tourists but gold mining has ceased. [9]

Aerial photographs show the Ralston Mine on Auraria Road near the intersection of Harms Road and Ben Higgins Road. Auraria Road may include the old trail called Gold Diggers Road that linked Auraria and Dahlonega. [10]

Gold from the Ralston Mine was shipped to the Philadelphia Mint in the "glory days" of Lumpkin County. A tall brick U. S. Mint was built in Dahlonega in 1838, and this mint processed the Ralston Mine gold. The mint closed with the beginning of the Civil War in 1861.

Lewis Ralston had protected his farm through all the trauma of the Removal in the years 1838-39. He lost his gold mine, but he may have worked for other mine owners or he may have concentrated his efforts on improving his farm on the banks of the Chestatee River. He had lost many of his other properties through the Land Lottery. There may still have been need for his ferry operation. He may have raised cattle for the Atlanta market, or he and his family may have cultivated a field of corn. He did not try to raise cotton. In 1818, cotton sold for 30 cents a pound, but by 1840 the price had dropped to only 4 cents for each laboriously grown pound. (10)

The Ralstons were part of the group that read a letter from Jenny Wimmer, a former Lumpkin County woman, who had traveled to the West with her husband. Jenny urged the gold miners of Lumpkin County to come to California – and early in 1849 some of them did. The Ralston family, Elizabeth and 13 children, discussed how they might finance a trip for father Lewis. The expense of a prospecting outfit for him would be a major investment. The two oldest sons, John, 23, and Lewis, 13, would help Elizabeth. Lewis knew he was an experienced miner and he was determined to succeed. The story of his 1850 prospecting trip follows in Chapter V.

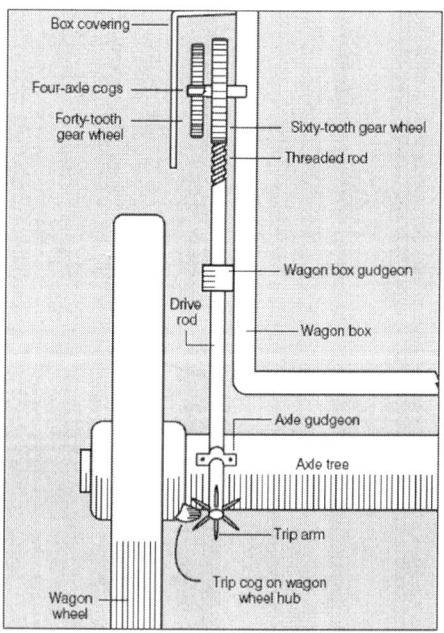

Odometer – fixed to wagon wheel to measure distance.

CHAPTER V

⚜ LEWIS RALSTON & COLORADO GOLD ⚜
1850

Elizabeth Ralston welcomed letters from her family in Indian Territory so many miles away. She remained especially close to her cousin Emily Duncan who had married a Baptist minister named John Beck. Reverend Beck also had Cherokee ancestors. Emily no doubt sent Elizabeth clippings from the Cherokee newspaper *The Cherokee Phoenix,* which on January 8, 1849, told of the discovery of gold in California. This newspaper, originally published in New Echota "back home" in Georgia, was now written and distributed in the new capital of the Cherokee Nation, the town of Tahlequah, Indian Territory.

In the early months of 1850, Pastor John Beck, gathered interested men of the Nation together and encouraged them to think of the wealth that could be theirs for the taking. A prospecting party was organized. Emily Duncan Beck wrote to the Ralstons about the proposed gold seeking venture. In Lumpkin County, Clement Vann McNair was the leader of a similar group. The McNair party including Lewis Ralston soon moved out, bound for Tahlequah. No records have been found for the organization of the McNair party.

Patricia and Dr. Jack Fletcher have gathered documents regarding eleven different small groups of men that started their westward bound prospecting journey from the great salt spring near Tahlequah, Oklahoma. These original groups met, established "rules for the road" and eventually settled into four wagon companies. The Fletchers and Lee Whitley combined many sources in their massive book *Cherokee Trail Diaries.* Their assembled material traces the groups as they divided, regrouped or merged along the trail. The year before, in 1849, the fear of Indians and the vastness of the plains ensured that wagon trains would agree to cling together for mutual benefit. That was not true in 1850. (1)

All of the 1850 groups leaving from Cherokee Nation West (Oklahoma) followed a trail that was established by wagon master Captain Louis L.

Evans, honored as the "Father of the Cherokee Trail." In 1849 Evans led forty wagons and 130 prospectors safely to the California "diggings." He pioneered a route west along the Arkansas River, north to the Oregon Trail, then west again through what would be named Wyoming. The Oregon Trail, a rugged trace carved by the feet of fur trappers, covered 2,000 miles from Missouri to the mouth of the Columbia River. (2)

Louis Evans, with his father and brothers, had founded the small village of Evansville, Arkansas. The Evans men were together on the 1849 trip. Louis Evans wrote a good account of his journey and his notes were hand copied and passed along to other wagon masters. He marked an important crossing with a message on a large rock which served as a guide post for future groups. Here at the confluence of Turkey Creek and the Little Arkansas River, the rock marked the cut-off for the old Santa Fe Trail pioneered by traders in 1811. In selecting the route for his 1849 prospecting party, Evans followed old Indian trails or the footpaths made by fur trappers and traders who, in the 1830s, walked the dusty paths carrying large back packs of beaver skins from Fort Laramie (Wyoming Territory) to Taos, (later New Mexico). Whiteley, a partner of the Fletchers, quotes John Pyeatt who mentioned an "old trail" followed by the Evans party.

Unfortunately, the Evans journal has been lost. Alfred Oliver had possession of the Evans diary and supplied information to other wagon trains. Exhaustive research by the Oliver family and the Fletchers has been unsuccessful. Louis Evans chose to assemble his group at the Grand Saline, a practice adopted by other wagon trains. (3)

In 1850, some of the captains of the four main wagon trains kept records or had people who wrote journals of the trip. All met near "the Grand Saline" as the point of beginning. Lewis Ralston and the group from Lumpkin County would join a group which included some Cherokee, some members who were partially Cherokee, and some Anglo-Saxons - gold seekers all. A young Cherokee boy named John Lowery Brown was part of this group and he carried a quill pen, a bottle of ink and sheets of paper on which to write an account of his adventure. After some days on the trail he purchased a small notebook and his record today is known today as the John Lowery Brown Diary. The group elected Clement Vann McNair as Captain, but because Brown's writing was so detailed, this 1850 group sometimes has been referred to as "the Brown wagon train."

There are no organizational records for this group but they were probably similar to those adopted by the Cherokee Nation Emigrating wagon train:

Tahlequah, February 3, 1849

> "Whereas the intelligence which has reached us of the California gold mines is corraborated (sic) by official reports, which render it certain that there is sufficient (gold) for all who wish to avail themselves of the opportunity of improving their fortunes . ." (4)

After further lengthy information the document asks the members to subscribe to the following conditions. They would agree to stay together, they would recruit at least 100 members, everyone would be armed with "a rifle gun, a butcher knife. . . and not less than 3 lbs of powder and 9 lbs of lead." Each prospector would also promise to provide themselves with "100 lbs of bacon, 200 lbs of flour, 25 lbs of salt, and 2 lbs of soap." Everyone would be required to have a wagon pulled by cattle (oxen) or mules, and each wagon would contain 6 gallons of tar, 1 handsaw, 1 drawing knife, plus augers and chisels. Each wagon would be subject to examination by a committee.

In addition to supplies similar to those listed, almost every wagon train included those who also provided themselves with some rudimentary medicine. Some trains, as did the McNair group, promised to pay a doctor with some of the hoped-for gold if he would travel with them.

A medicine chest might include the following:

Laudunum (opium)	Peppermint
Cayenne pepper	Calumel
Physic	Castor oil
Sweet oil	Whiskey – small bottle

The Grand Saline was known by early travelers as the best place to ford the Grand (Neosho) River at its confluence with the Saline River. Near the Grand River was a "salt-laden" spring. The spring and surrounding marsh were a truly commercial advantage in the new land. Chief John Ross used Removal funds to purchase the area and established his brother Lewis Ross as operator of the salt works. Lewis Ross soon opened a general store on the east side of the Grand River and his son, Robert, received a patent for a U. S. Post Office. The salt water was "boiled down" yielding about "15 bushels of salt a day." True to Cherokee laws, the profits from the operation were used by the Nation for the benefit of all enrolled members. (5)

The Grand Saline was a popular rendezvous site. It was only 22 miles north of the new Cherokee capital at Tahlequah. At the Ross store, last-minute supplies could be purchased and final letters mailed home. Surrounding fields provided ample space for the oxen, pack animals and carts of the eager gold seekers.

On April 9, 1850, the Cane Hill Emigration Party was the first wagon train to leave the Saline for the distant gold fields of California.

On April 20, 1850, eleven days later, the group headed temporarily by Rev. Samuel Houston Mayes, including Lewis Ralston, headed north from the spring and the Neosho (Grand) River. Either before or after the Cane Hill group and before the Mayes group, a wagon train headed by "Captain Edmonson," later mentioned by diarist Brown on June 20, also took to the trail from the Grand Saline. Although Brown recorded the Edmonson party, no other record has been found of the Edmonson (or Edmundsun) wagon train. Also, according to the Fletchers, there was another four-wagon group of Cherokees headed by Thomas Fox Taylor and Devereaux Bell that later united with the Mayes wagon train. (6)

Thomas Fox Taylor was the grandson of Charles Fox Taylor, long a revered as a Cherokee leader. Thomas attended Knoxville College in Tennessee and became President of the Cherokee National Committee. He was known as a great public speaker and worked with Chief John Ross to try to avert the Removal of the Nation to Oklahoma. (7)

On May 5th, the Mayes party caught up with the group headed by Clement Vann McNair from Georgia, and by this time the group also included the wagon train headed by Thomas Fox Taylor. On this date "Clem" was elected as captain of the group.

The McNair and Mayes parties, both from Georgia had an easy relation-ship. The party headed by Taylor, however, did not agree with some of McNair's rules. Even though McNair was elected to be the captain of the combined wagon train, the Taylor party brought dissension.

At the Grand Saline the McNair train had waited for the group from Tahlequah that had been organized in part by Rev. John Beck. They finally moved out and the Beck party joined them at a point on the trail near today's east-central Kansas. On May 5, Brown wrote in his diary: "105 men, 15 negroes (sic) and 12 females all under the command of Clem McNair." (8)

Clement Vann McNair was the youngest child of David and Delilah Vann McNair. Delilah was an accomplished woman of Cherokee heritage. Her father, although not a chief, was an influential leader. David McNair "had a beautiful white house . . . and six or seven hundred acres of the best land . . . he raised about five thousand bushels of corn (in one year)." Clement was well educated and was a delegate from the Cherokee Nation to Washington D. C. in 1846. His family lost everything in the relocation of the Cherokees. (9)

John Lowery Brown started making notes for his journal in April. Seemingly he purchased a small book later in the journey and transferred his scattered notes to it. He continued making an entry every day. On April 23, 1850, Brown traveled to "Mothers", some miles from his own residence nearer Tahlequah. His reference was to his own mother Rachel Lowery Brown. Rachel Lowery was a descendant of Chief John Benge and related to Lewis Ralston's wife. She had married David Brown. Her father was George Lowery who assisted John Ross as co-captain (with John Benge) of the first contingent in the Removal, the first group leaving Georgia for new Cherokee Nation West. George Lowery, the diary keeper's grandfather, had helped with the first printing of Sequoyah's alphabet and had served as Assistant Principal Chief of the Cherokee Nation in 1847.

The McNair wagon train began rolling west on April 28. On May 15 the party camped at a place recorded by Evans as "Buffalo Chip Campground." For the first time there was no wood and to their surprise the buffalo chips worked well as fuel for a fire. May 20 Brown noted that they had "20 waggons (sic) and 1 carryall." On May 23 they passed "a large heard (sic) of buffalo." On May 31 they met Major Fitzpatrick at the Lower Crossing of the Arkansas River – present day Cimarron, Gray County, Kansas.

Federal Indian Agent Major Thomas Fitzpatrick and his staff were surrounded by a great crowd of Indians. There were five major Indian tribes in attendance. In addition to the annuities that Fitzpatrick was authorized to dispense, brisk trading was carried on between men of the McNair party and the white traders traveling with Fitzpatrick. It is here that diarist Brown probably purchased his little book. (10)

John Lowery Brown's small diary measures 4 by 6 inches, bound with a cardboard cover. A tan leather-like fabric is glued to the cover. It contains 75 unlined pages. Brown used a quill pen dipped in his ink bottle to make his entries. The author held this small powerful journal in her hands when she visited the Gilcrease Museum. (11)

Brown wrote "Off for California" on the first page of his journal. California would not be granted statehood for five months, not until September 9, 1850.

Each day Brown noted the date, the number of the camp (excepting the first few days), and the miles traveled. The mileage is remarkably accurate. The group may have had an odometer attached to one of the wagon wheels. His handwriting and spelling compare most favorably to other documents from the 1800s.

There is no list of the names of the wagon train members. The names he does include are listed at the end of this chapter. Brown's account is one of the best of the gold rush diaries.

Members of the wagon train fell silent at the first sight, June 19, of the once majestic adobe castle built by William and Charles Bent. Bent's Fort combined the security of protection with a hotel and trading post. It had opened in 1833 with a license acquired from the federal government. Sixteen years later, an angry William had rolled in casks of "powder," with his own hands, and blown up his palace. The date was August 21, 1849. He was bitter at the U. S. government's use of his fort as a barracks for the army and the uncaring officials who refused to pay for the use. Bent had built a new smaller fort at Big Timbers on the Purgatoire River. This compound was later taken up by the government and renamed Fort Wise, and later renamed Fort Lyon. He refused to leave his masterpiece to the army with no credit or payment and cringed at the thought of Indians living in his fort with their ponies, bundles and lively ceremonies. (12)

The McNair train did not stop at the ruins of Bent's Fort. On June 12 they marveled at the tops of snow-covered peaks in the distance and reached Fort Pueblo on June 13. They had a "lay-bye" the next day including a "big dance" that night. Frances Parkman described the fort in 1846:

". . . nothing more than a large square enclosure surrounded by a wall of mud miserably cracked and dilapidated." (13)

The group traveling with the McNair party, but taking their orders from Thomas Fox Taylor, sold their wagons and purchased mules and pack saddles upon reaching crumbling Fort Pueblo. They were ready to push ahead of McNair's slow moving horse and mule wagons, and they refused to waste time in worship that required a stop every Sunday. The McNair party did enjoy the Sunday "lay bye." Taylor's party asked for travel to continue every day to reach the gold fields in record time. Captain McNair said, "no," and the Taylor group quickly moved out.

Early on the 14th, McNair's train moved north along the Fountain River, killing a bear on June 16, a fine addition to their food supply. The next day they passed "Pikes Peak covered in snow" and the following day they traveled beside a sparkling stream named Cherry Creek because of the wild cherry bushes growing on the creek banks.

John Lowery Brown's Journal (as written by him):

June 17   Traveled north, leaving the Creek. Traveled Over Sand hills, pine Timber. Passed Pikes Peak which is covered with snow. Camped at cold spring of water – made today about 20 miles. Camp 40th
(near present day Colorado Springs)

June 18   Traveled 25 miles. Camped on Bold Running Clear stream of water waters Of the Platt. Good grass & wood.
Camp 41
(In the left margin) today we crossed the dividing ridge between the Arks & Platt

June 19   Continued down the above mentioned Creek 20 miles Good grass water & Timber Camp 42
(In the left margin) very hard storm this evening hale from the size of a Birds to a hens egg.

(The McNair wagons were rolling north on Cherry Creek approaching its confluence with the South Platte River.)

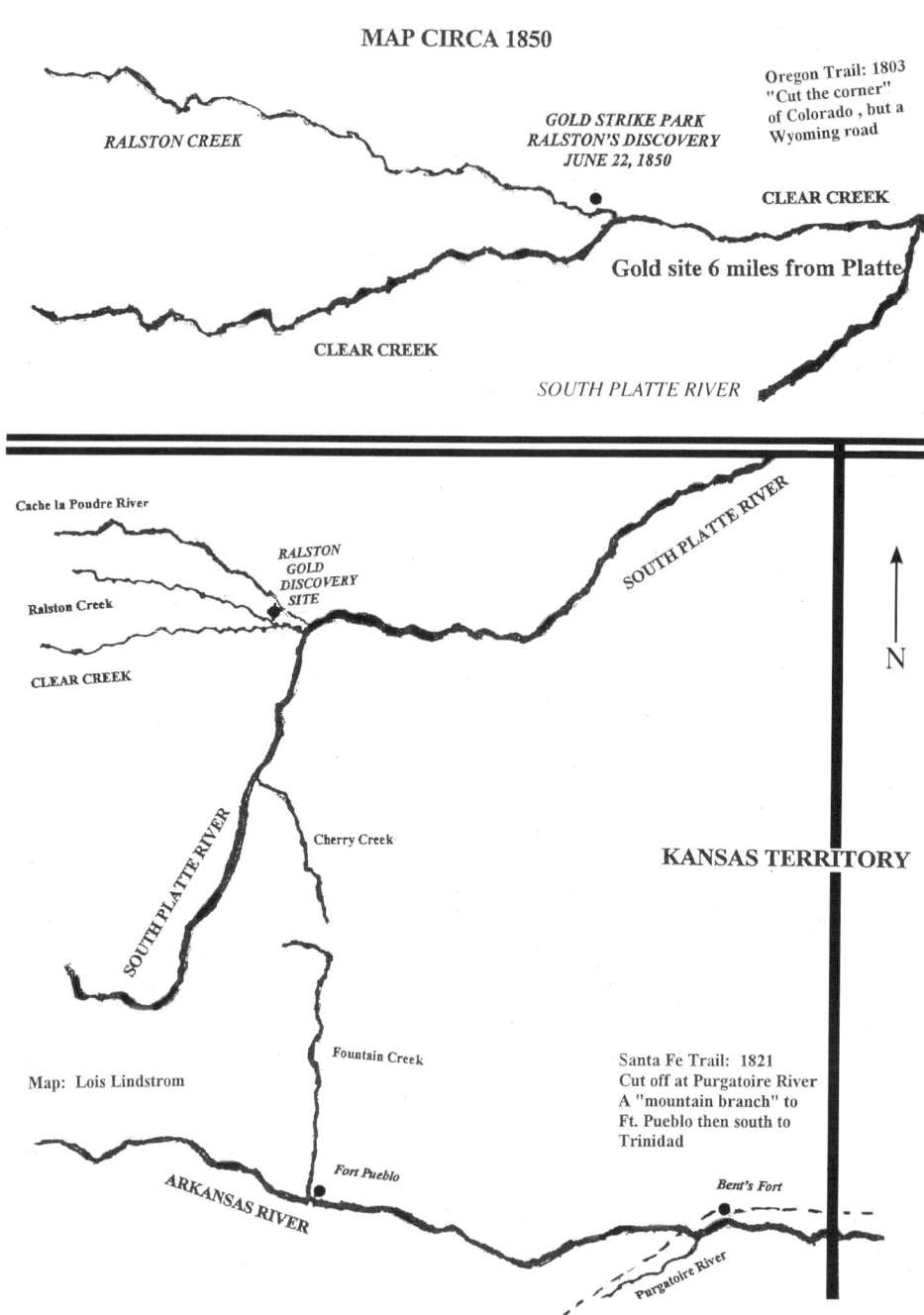

MAP CIRCA 1850

RALSTON CREEK

*GOLD STRIKE PARK
RALSTON'S DISCOVERY
JUNE 22, 1850*

Oregon Trail: 1803
"Cut the corner"
of Colorado , but a
Wyoming road

CLEAR CREEK

Gold site 6 miles from Platte

CLEAR CREEK

SOUTH PLATTE RIVER

Cache la Poudre River

*RALSTON
GOLD
DISCOVERY
SITE*

Ralston Creek

SOUTH PLATTE RIVER

CLEAR CREEK

N

SOUTH PLATTE RIVER

Cherry Creek

KANSAS TERRITORY

Fountain Creek

Map: Lois Lindstrom

Santa Fe Trail: 1821
Cut off at Purgatoire River
A "mountain branch" to
Ft. Pueblo then south to
Trinidad

Fort Pueblo

Bent's Fort

ARKANSAS RIVER

Purgatoire River

June 20	Took a left hand trail down the creek which was made by Capt Edmonson about two weeks ago. About 10 oclock came to the South Fork of Platt River. Made a Raft and continued crossing the wagons camped on the Bank of Platte. Camp 43

(There are no existing records for Capt. Edmonson. What Brown calls the "South Fork" was Vasquez Creek. It was named by Lewis Vasquez, a fur trader of the 1830s who erected a small structure, Fort Convenience, at the confluence of Vasquez/Clear Creek and the Platte River. The diary keepers of 1850 and 1858 did not mention the ruins of "Fort Convenience" and the simple structure had evidently succumbed to the ravages of time.

McNair's wagon train did not continue their advance to the confluence of Cherry Creek and the South Platte River. Scouts riding ahead of the wagon train reported the banks of the Platte were overflowing with a muddy deluge. Waters of the mighty river were rushing from melting snow of the high peaks to the dry plains of the east. McNair ordered his group to bend their wagons east, following the riverbanks at a distance, looking for a safe place to cross. They all knew the peril of trying to ford a flooding river. They reached the confluence with the Platte River and Vazquez Creek. (It would be renamed Clear Creek.) Brown called the Vasquez the "south fork."

The weary men, using hand axes, felled trees and made at least one raft. By nightfall most of the party had safely crossed and they camped on the west side of the Platte River - the north side of Clear Creek. The crossing was finished the next day and Brown had time to write a description of their arduous labors:

"Saying of the boys while wrafting the Platt – No one speak but the Captain – Will you hold your tongue you scoundrel – hold on pitch on to that raft fellows a dozen or two of you – push it off – now she rides – Let her swing – hold to the rope to the right you Rogues – run out to the right with the rope . . . now she Rides – Get off the rope there behind – I can't pull the rope and you on the Rope" . . . (and much more).

June 21	finished crossing at 2 o clock left the Platt and traveled 6 miles to Creek. Good water grass and timber Camp 44
	(in the left margin- author's italics)
	we called this Ralstons Creek because a man of that name found gold here.
	(opposite June 22 entry)

This simple entry establishes the Ralston find as the first documented gold strike in Colorado. It was placed on the State Register of Historic Properties on December 1, 1995. There had been other random gold finds, but no one marked the spot or left directions needed to find the place of the gold discovery. No other early gold find in the Rocky Mountains prompted a following prospecting party to return to the site.

Prospectors returning to "Ralston's Creek" became the settlers who would be the spur to the founding of Arvada 20 years later. No other journal for 1849 or 1850 spoke of gold panning along the trail. Reverend John Beck said Lewis had panned gold worth $5 and this small discovery would open the door to the Colorado Gold Rush of 1859.

Recognition of the Ralston find was 145 years in the making. This was due in part to the fact that Brown's diary was lost and thus unknown to early Colorado historians, and also because the Ralston gold site could not easily be identified. The author marked miles on a cord and traced all possible routes to Ralston Creek on a map, following the mileage listed in the Brown diary. The above account is the only correct conclusion. The most important clue is the phrase in Brown's journal: "traveled 6 miles to a creek." The creek of course was Ralston Creek and it joins Clear Creek six miles from the South Platte River.

On that momentous day, June 22, 1850, the McNair wagon train was electrified by Lewis Ralston's shouts of "Gold!"

| June 22 | Lay Bye. Gold found. |

| June 23 | this morning all except 3 messes who traveled on Concluded to stay and examine the Gold. Bell, Dobkins and R. J. Meigs traveled on. |

| June 24 | Left Ralstons Creek and made 26 miles. Rainy and very muddy. Camped on creek plenty water wood and grass. Camp 45 |

The McNair wagon train had spent at the most little more than two days on Ralston's Creek. Although members of the wagon train quickly began panning for gold in Ralston and Clear Creeks, Brown does not record their finds. Ralston had $5 worth of gold in his pan. In 1850 that was equal to about one-third of a troy ounce. A full troy ounce was worth $16.00. (14)

Lewis Ralston's name is mentioned only once in John Brown's diary – that mention occurs on the day he discovered gold, June 22, 1850.

The men were in good spirits as they moved north toward the Oregon Trail. Brown mentions "good water." The train caught up with those who had rushed ahead, Bell, Dobkins and Meigs, at the Cache la Poudre River near present-day Fort Collins. On June 28, the Charles Fox Taylor group rejoined the McNair wagon train.

The prospectors did not realize that the hardest part of the journey was still ahead. The challenges of fording even larger rivers, of trudging up and down steep mountains, of crossing burning deserts, of suffering from the lack of water and food would test even the strongest man. They all, including Lewis Ralston, were ready.

Names that are mentioned in the Brown Diary:

Adair, George Washington

Barker, Dr.

Bean, Tolbert
Died September 6

Rev. John Beck
Baptist minister

Belle, Devereaux J.
Also spelled Bell

Belle, Juliette
Wife of Devereaux

Brewer, Oliver Hazard
Called Perry

Brown, John Lowery

Bushyhead, Dennis W.
*Not mentioned in diary
but was with the group*

Clark, John

Davis, ---------
Died August 17

Dobkins, -----

Emeg, --------
Killed by Indians
August 24

Estelle, J. M.

Fields, -----

Henry, Archer

Hildebrand, Jack
Died August 11

Home or Holmes

Huffaker, John A.
Died December 11

Hunter J. B.

Jonas – black boy
Died September 14

May, J. J.

Martin, George M.

Mayes, Samuel Houston

McDaniel, Charles

McNair, Clement Vann

Meigs, Return Jonathon

Mims, T. J.

Oliver, -------

Pack, ------

Palmer, -----
Died August 4

Palmer, Dr.

Ralston, Lewis

Russell, ------
Died August 7

Simons, Samuel, *perhaps
Lewis Ralston's brother-in-law*

Street, Henry

Taylor, Thomas Fox

Tuff, Runaway

Trott, Ben F

Wolfe, John H.

CHAPTER VI

∼ⓔ CALIFORNIA GOLD ℗∼

Primary source for the following is the John Lowery Brown
diary with his spelling and punctuation.
Some of his daily entries have been condensed.
Muriel H. Wright's typescript of the diary is very helpful.

The McNair wagon train left Ralston Creek, heartened by the discovery of gold and traveling hard to reach the three wagons who had traveled north ahead of them. They put the heated discussion about "stay" or "go on" behind them. They were determined to reach the golden bonanza in what, in only 70 more days, would be named the state of California. The united group built a raft to cross the Big Thompson River, then soon had to cut trees for a second raft to cross the Cache La Poudre River (present-day Fort Collins.) The Taylor "pack" group joined them at this camp and again caused dissension with their request that the practice of no travel on the Sabbath be abolished. They urged a plan for faster travel and an earlier arrival at the "diggings." Clement Vann McNair resigned on June 28 and Brown wrote: "T. F. Taylor the Lieutenant took command of the Co. as Captain." On June 30 they crossed what would be the future border of Wyoming and began following the tracks of "Captain Edmonson's ox train."

July 4 was notable for the sweet-smelling sage they found to fuel their campfires, but later that night "about 25 head of horses & mules were stolen from our Ca. by Indians." Captain Taylor and a few other volunteers immediately began tracking the hoof prints and did not return to the moving train until July 7. They failed to retrieve a single animal.

They reached "Sulphur Springs, not fit for man or beast to drink" on July 13, and the men and animals who tried to drink the water almost immediately suffered cramping illness. The party struggled onward, but by July 20, Captain Taylor had persuaded the former McNair party to abandon their wagons and "pack." Food and a few cherished items, including Brown's diary, ink and pen, were wrapped in bedrolls, and the packages were tied to the backs of horses and mules. The gold seekers walked.

They crossed into what would become Utah and Brown wrote: "we find this trace to be crowded with Emegrants to the Gold diggins. We are hardly ever out of sight of waggons." The livestock of earlier travelers had eaten all the grass and the water holes were trampled and muddy. On July 29 they paid a toll to use the "Mormon bridge" and on July 31 they tramped through "Mormon City" (Salt Lake City) and camped on the banks of the Jordan River.

The weary men felt the worst was over not realizing that the rutted track would lead them through one dreadful desert after another. They chose to follow the Hastings Cutoff (avoided by most wagons trains due to the Donner tragedy) which curved around the south end of the lake and then west on the Donner trail.

The well-provisioned Donner party had traveled west in 1846, planning to establish a ranch in Mexican Territory (California). They started late in the year, traveled at a leisurely pace and stopped to visit friends and historic sites along the way. They reached the Sierra Nevada mountains in the fall where they were halted by an unexpected heavy early snow. They were only 22 miles from the summit, a hard climb through the first snow drifts, but the downward trail would have brought them to a warm valley. They decided to stop and "wait out" the storm. They were buried by the massive snowfall and from the party of 89 travelers, only 47 survived. Their history is marred by the suggestion of cannibalism.

Captain Taylor knew the Donner story and he encouraged and prodded his group. They finally reached Hope Wells (Elbow Springs) on August 7. The oasis was crowded with prospectors, many of whom were dying from cholera. Taylor's weary party stopped for a "lay bye" and eight members of the McNair/Taylor party expired within a few days of each other. Dr. Barker from Missouri, with a small party of eight men, agreed to join the Taylor party, bringing the total number of their party to "53 persons."

The original McNair group had numbered 132 people on June 22. Brown did not list the names of those who joined or left their group. He did try to list the names of those who died on the trail. By August their group was half the size it had been a just a little more than one month earlier. Some may not have died – they may have left the wagon train at Fort Bridger or Salt Lake City.

After a day of rest at Hope Wells the McNair/Taylor company crossed yet another desert to reach water, another small "hole," this one also named Sulphur Springs. On August 8 and 9 they rested during the day, and traveled the desert at night, finally reaching Relief Springs on August 10. Brown wrote that they had traveled 45 miles "without stopping and without water for our horses."

At Relief Springs former Captain McNair and others were too ill to travel so Dr. Barker remained in the camp with them. On August 10, the small group crossed yet another desert and reached the foothills of the Sierra Nevada mountains. Here they found green grass and pure water. Brown wrote: "quite a God send." Here they were joined by McNair, Dr. Barker and others. Captain Taylor wanted to abandon rest and to travel on immediately. The group rebelled against Taylor's orders. "C. V. McNair's, Mayes and Martin's messes . . . and Dr. Barker & his men separated themselves from Capt. Taylors Ca." Brown stayed with Captain Taylor, and they, that day, marched resolutely down the banks of St. Mary's River, later named the Humboldt River.

On September 5, his hand trembling with fatigue and hunger, Brown wrote: "We have had no bread since Aug. 28 . . . no flour to be had for love or money." Taylor's men were starving. Again they were faced with cutting grass and packing it for their animals. Two days later they reached Carson Creek where there were traders willing to sell food. Brown wrote: "Great many folks here, great many dying. Camp 105."

Jonas, a trusted slave and willing worker, died on September 14. He was owned by Mrs. Elizabeth Pack, related by marriage to diary-keeper John Brown. Mrs. Pack lived near Tahlequah and had sent Jonas west to possible freedom, trusting him to John Brown's care. California had been admitted to the Union on September 8, 1850, as a "free" state. Jonas, whether he knew it or not, had been a free man for seven days when he died.

The Oliver train, 5 miles behind Taylor's group, had to "lay bye" a day while Barbara Hildebrand Longknife gave birth to a baby girl. Several of the nearby gold seeking parties rejoiced when they heard the news.

On September 19, the small group of McNair/Taylor survivors crossed the summit of the mountain "over the worst road I ever traveled." (Carson Pass south of Lake Tahoe). Brown had heard of the pass and wrote about it on September 16: "At this place & up this Stream there is a pack rout across the mountains which is said to be the nearest though the roughest way, than the waggon Road."

They descended into a lush California valley and learned that they were approximately 100 miles distant from Sutter's Fort. (The fort, built by Johan Augustus Sutter, became the nucleus for Sacramento City.) The Taylor pack train began to come apart, as single men or small groups split away to explore alone. Brown's horses were stolen and members of the Adair and Mayes families put his packs on their animals and they all walked. Brown listed the following gold camps as they trudged on:

Leak Springs	El Dorado County
Camp Creek	30 miles east of Placerville
Pleasant Valley	10 miles southeast of Placerville
Ring Gold	On the road south of Placerville
Weberville	5 miles south of Coloma
	(Locations by W. O. Waters) [1]

On September 28 they reached Dead Man's Hollow. They were only five miles from Coloma where James Marshall had found the famous gold nuggets. They began building log shelters and on November 2 Brown wrote: ". . . my mess moved into our Cabbins which was the first time that I slept in a house since the night of the 27th of April." Aside from a few other brief comments, the building of the log house marks the end of Brown's diary.

We do not know the other individuals in Brown's "mess." (Mess was the term for a group that ate and slept together – both on the trail and in a dwelling.) George Washington Adair, plus Reverend Samuel Houston Mayes and his two sons, George and William, built the cabin next to John Brown's. Two Coffinberry brothers were the official census takers for the 1850 census. They reached Dead Man's Hollow in October of 1850, before the cabins were finished. They did not list either John Brown or Reverend Mayes. They did list the Mayes brothers and G. W. Adair. California census records for 1850 do not list Lewis Ralston. The Coffinberry brothers tramped the major hills and vales of El Dorado County, but there were individual miners and small groups in every ravine and the census record does not list them all. [2]

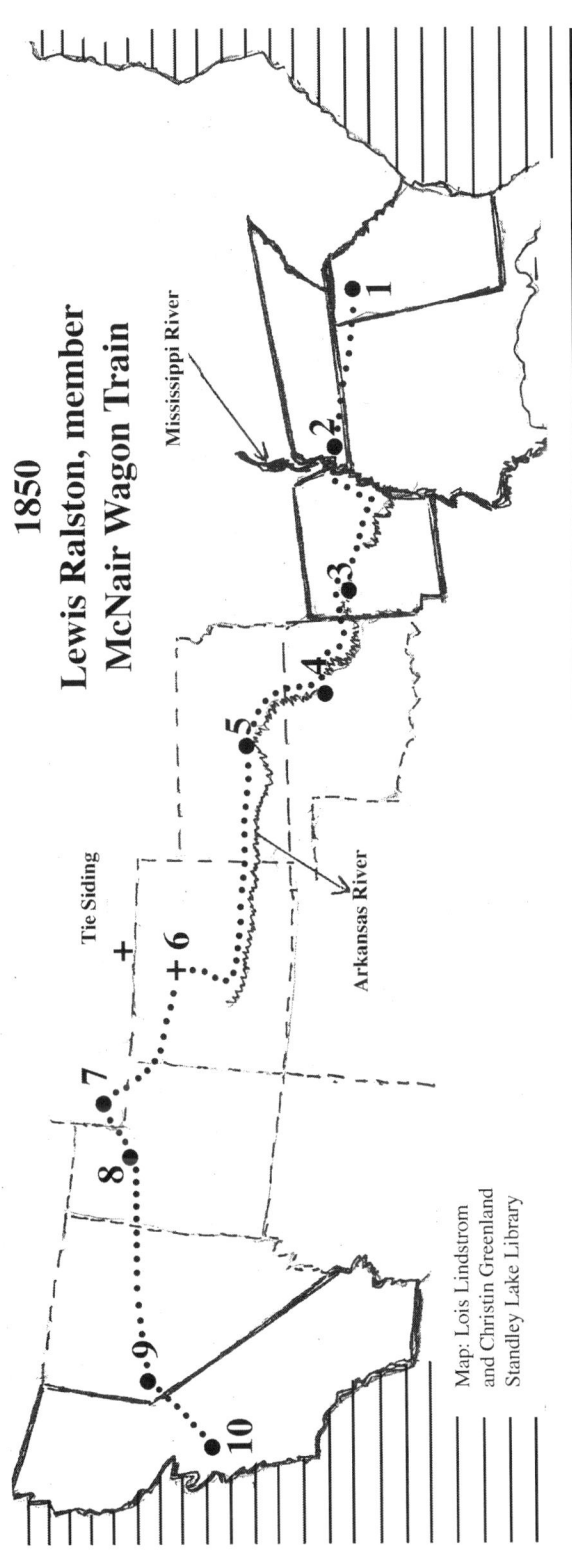

1850
Lewis Ralston, member
McNair Wagon Train

Mississippi River

Tie Siding

Arkansas River

Map: Lois Lindstrom
and Christin Greenland
Standley Lake Library

Location	Statehood
1 Dahlonega, Georgia	Jan. 1, 1788
2 Memphis, Tennessee	Jan. 1, 1796
3 Fayetteville, Arkansas	June 16, 1837
4 Tahlequa, Oklahoma	
5 Deerfield, Kansas	
6 Arvada, Colorado	
Gold found June 22, 1850	
Founded 1870	
Statehood 1876	

Location	Statehood
7 Bridger, Wyoming*	
8 Salt Lake City, Utah	
9 S. Lake Tahoe, Nevada	
10 Carson Pass to	
Coloma, California	September 8, 1850
* Colorado to Wyoming	
through Tie Siding	

The wagon train traveled shore to shore through what would become
ten states. Statehood had been achieved by only four in 1850.

55

The gold mining town of Coloma, the home of the first gold seekers of 1848-49, grew on the southern bank of the river, opposite Sutter's sawmill on the South Fork of the American River. The valley was first named "Colloma" (peaceful) and was inhabited by gentle Indians. Captain Sutter befriended them, and hired them to work at his fort. The mining camp was granted a United States Post Office on January 13, 1851 under the name Coloma, and the spelling was used from that time forward.

Coloma was the most important town in the northern gold region, second only to Sacramento City which developed (1848-49) at the confluence of the American and Sacramento Rivers. Here Sutter had built his fort. Sacramento later was named the capital city of the state.

The area of Coloma, which included the first gold mines, became the County of El Dorado. (Dorado is the name for a constellation of glowing stars in the skies of the Southern Hemisphere – a fitting name for one of California's richest gold mining counties.)

The South Fork of the American River was a major water source, a rushing torrent between steep banks. A ferry service over the river developed in 1848 - perhaps used by the surviving men of the McNair/Taylor prospecting party. (3)

Coloma's log houses clustered at the base of a tall mountain named Mount Murphy. The mountain was named for an early arrival, an Irish stonemason called Patrick O'Brien Murphy. He built a stone fort on the mountaintop as protection from the Indians, even though there is no record of Indian troubles in the Coloma valley. Murphy borrowed or purchased a large cannon from Captain Sutter. He fired the cannon whenever he saw the approach of the mail-bearing stagecoach from San Francisco. The sound boomed across the valleys and the prospectors all rushed to town to receive their letters. (4)

There is no accurate account of the numbers of gold seekers that flocked to California in the early years. Some writers estimate 75,000 which may be too small a number. By 1850, the late-comers found many abandoned log shacks, surrounded by the debris of careless men, mole hills of dirt over small shafts and hills denuded of trees. (5)

There is no accurate account of the gold removed by these early efforts. There were three major gold mining areas in the state, and pockets of gold could be found on many small streams. An 1849 report stated that $10 to $40 million in gold had been mined, but not until 1852 was a more accurate figure reported that showed $81 million for that year. (6)

By November of 1850 very few prospectors were using only a gold pan. Men had banded together, pledging an honest division of treasure for all members of the group. Brown wrote about "throwing up dirt." Soil was dug and shoveled into a "rocker" which was like a giant flour sifter. If the yield was promising, a "long tom" was built. This device was a long wooden open box with riffles across the bottom at regular intervals. Dirt was fed into the box, and water from a stream was diverted into the sluice box. As the soil was "washed" the heavy gold sank to the bottom of the "tom" and caught on the riffles. (7)

California was originally Mexican Territory. United States troops marched to a territorial war against Mexico in the years 1846-48. The Mexican War ended with a treaty on February 2, 1848 and the area that would be named California was ceded to the United States. Less than a month before the treaty was signed (16 days), James Marshall on January 14, 1848, made his discovery of gold at Sutter's Mill. Mexican officials certainly would have been less eager to give up the land had they known about the gold. Officers of the U. S. Army served as officers of the new California Territory.

Captain Johan Augustus Sutter, born in Switzerland, was a Mexican citizen. He had official papers from Mexico City granting him the right to build his fort and colonize the area. Sutter landed on the shores of Monterey Bay in 1839. His land grant extended over 50,000 acres and his fort became a trade center for the area. He employed Mexican laborers and rescued many immigrants. The latter he helped learn an employable trade such as blacksmithing or the tanning of hides. He encouraged these newcomers to purchase land from him and establish a farm – all subject to his require-ments. His colony grew – it would include 200 families by 1850. Finished lumber was necessary for home building and in December of 1848, Sutter sent James Marshall and others into the hills to build a sawmill on the South Fork of the American River. Peter and Jenny Wimmer were part of the crew - Jenny was the camp cook and Peter was a skilled carpenter.

When Marshall found his golden nuggets, "half the size and shape of a pea" he asked Jenny to boil them in her soap pot. In addition to cooking she was also the official washerwoman and made her own soap from fats, ashes and lye. After the overnight soak in the soap, Jenny pulled the gleaming gold nuggets from her pot and Marshall wrapped them in his handkerchief and galloped off to the fort. Sutter was horrified and tried, in vain, to stifle the news. It spread like morning sunlight. In a few days some of Sutter's men, Mormons and Mexican laborers who were building a flour mill in the area, threw down their tools and rushed to the site of the sawmill. Jenny wrote a letter to folks back home in Georgia.

The news spread to Sacramento City. Sutter could not stem the tide of wild-eyed gold seekers. They tramped throughout his domain, neither recognizing nor caring that they were trespassing. In the years to come Sutter tried every tactic to try to protect his land and to collect damages from either the gold seekers or the United States government. His efforts failed and he died a pauper. [8]

Prospectors who were at all fortunate might find gold so fine it was called gold dust. A "pinch" of this dust could buy an apple, a fresh green onion or an alcoholic drink. Miners carried such dust in a glass vial if they had such an item, or in the hollow stems of goose feathers. Some miners had special vests with tiny pockets. Others carried small leather bags or the hollow horns of animals. The Wells Fargo Stage Coach Company of San Francisco would accept dust and nuggets, and in return issue a check that could be mailed to families "back home." The San Francisco Mint opened in 1854.

James Marshall continued to live and work near Coloma. He owned a blacksmith shop, but often made a small profit from men who consulted him as to the best location where gold could be found. None of his ventures resulted in a personal bonanza. He died August 10, 1885, and was buried on Mount Murphy. His shop and cabin were sold to pay his funeral expenses. [9]

The 1850 diary-keeper John Lowery Brown returned to Tahlequah at an undetermined date. He was descended from the Benge and Lowery families and thus was related to Lewis Ralston's wife. His own wife was Anne E. Schrimsher. John placed the dairy in the secret drawer of an old desk.

Clement Vann McNair returned to Tahlequah, but soon, lured by the fertile soil, returned to California with his entire family. Thomas Fox Taylor stayed only a few months in California, returning to Cherokee Nation West where he became President of the Cherokee National Council. He died in 1862 fighting for the South in the Civil War. In 1851 Samuel Houston Mayes, who had returned to the Cherokee Nation, came back to California driving 200 cattle to the area of the gold mines – a gold find of another nature. George Washington Adair, a member of the McNair party, was killed fighting for the Confederacy in 1862. [10]

Did Lewis Ralston reach California? He cannot be found in any California census record or diary. Back in Lumpkin County, Georgia, the census taker reached the home of the Ralston's on September 24, 1850. Elizabeth and nine of her children were at home. Lewis Ralston's name, as head of the family, was listed first, a practice followed by many wives of absent prospectors. The census taker probably asked for names of all who were "in residence at this address." Birth of son Robert in 1854 would place him in Georgia by 1853. Elizabeth and the children, no doubt were glad to have him home.

Rocky Mountains
Kansas Territory 1854   Colorado Territory 1861
Photographer Joseph Collier
Courtesy Mary Collier Ross

CHAPTER VII

≈≈ TENSION & DEPRESSION ≈≈
1850 - 1857

The 1850 Federal census reveals little information about the Ralston family. What the census does show is the sad change in the family's financial situation. To begin with, Lewis, who had resided in the area for at least 25 years, and on previous census lists was called Lewis, in 1850 was listed as Luis Ralston. He gave his occupation as farmer and the amount of owned property was valued at $100.

Some of the children were living at home: [1]

Luis	46				
Elizabeth	38				
			Not listed		
Frances	31*	Rebecca born	1827	23
Elizabeth	18	John Tate	1828	22
Nancy	13	Alexander	1829	Infant death
Luis	13	Isaac	1846	Infant death
Louisa	10	James	1850	
Agnes	8	Robert	1854	
Henry	6	Martha	1858	
Amanda	2			
Zachariah	1			

*This age is obviously incorrect

In 1852 the Siler roll was initiated by the new headquarters of the Cherokee Nation West (Oklahoma) in an effort to list all members of the Nation who had remained in Georgia. (2)

The Lewis "Rolston" family was Number 17

Lewis	47	Listed as "white"
Elizabeth	39	"Mixed blood"
John	23	
Frances	20	
Elizabeth	18	
Nancy	16	Elizabeth and the children
Lewis Jr. *	14	were eligible for the
Louisa	11	Cherokee allotment.
Agnes	9	
Nancy	7	
Amanda	3	
Zachariah	2	* born 1837 so he was 15
James		

The two lists do not agree. The Siler list was very important because it was proof of blood relationship to the Cherokee Nation.

As is noted above, on the 1850 federal census, the $100 reported by Lewis for owned property reflects the downward spiral of the Ralston fortunes. There are records of court cases where Lewis Ralston's land was seized for non-payment of taxes. He had lost much of his property in the Georgia Land Lottery, the activity by which the State of Georgia took possession of Cherokee lands after the 1828 gold discovery. Ralston was not able to retain all of his property even though he was a white man, not a Cherokee.

By 1850 the economy of Lumpkin County was depressed and most of the Ralston's extended family and friends were gone, either in the 1839 Removal or the 1849-1850 California gold rush. Empty buildings lined the streets of Auraria. Dahlonega, the proud possessor of the post office, bank and courthouse was still a trading center, but the glory days of the gold bonanza were gone.

Auraria's only newspaper had moved to Dahlonega as had the post office. Historian Dr. Merton Coulter, in the following quotation indicates that Auraria had a postmaster, but this could not have been true after Dahlonega built the post office building in 1834.Coulter said concerning Auraria . . . " a small village of Lumpkin County. . . surrounded by a hilly region containing valuable gold mines. Auraria had five merchants including the postmaster."(3) George Paschal wrote to his mother in 1858 saying, "Auraria has fallen greatly into decay."(4)

Lewis and Elizabeth struggled to sustain their family. Lewis probably raised corn which could be fed to his farm animals. The animals and any surplus crops could be sold. A few deer still existed in the green forest, the former home of the Cherokee Nation, but Lewis would have been fined for hunting there.

The Ralston family may have attended the Antioch Baptist Church of Auraria. The church had been organized July 13, 1833 by Agnes Paschal, the wife of attorney George W. Paschal. When in 1841 Elizabeth gave birth to their ninth child, a girl, the couple named her Agnes Paschal Ralston in honor of the church founder. Early church records list only the names of officers, and the Ralston family does not appear in these early minutes. (5)

Neither Lewis Ralston nor Green Russell left a record of their first meeting. Green (Greeneberry) Russell and his brother John in 1849 had been part of the first California-bound wagon train to leave Lumpkin County. There is no diary or record of this party. They did not follow the same trail that was later used by the McNair/Ralston party. They probably used the Oregon Trail. Green and others returned to Georgia after a few years, traveling by boat from San Francisco, around the tip of South America, then north up the Atlantic coast.

Green used some of his new-found gold to purchase property on the Hightower River where he built a house for his mother and for his own wife and infant son. He organized a second prospecting party which included his brothers Oliver and Levi. The Russells were in California when the McNair prospecting group left for the gold fields in 1850.

Green's wife, Susan Willis Russell, like Elizabeth Kell Ralston, had Cherokee ancestors. The Russell and Ralston families may have met at Cherokee events. Lewis and Green may have discovered their mutual interest when they met while individually conducting routine business matters at the county courthouse in Dahlonega. Dr. Levi Russell, (according to historian LeRoy Hafen), Green's medical doctor brother, said they did not know of Ralston's discovery in the Rocky Mountains until they returned from their second California trip. Green listened and was greatly interested in the account of Lewis Ralston's discovery.

By 1850 railroads had been built west into the former lands of the Chero-kee. Georgia's first railroad had been chartered in 1832. Called the Georgia Road, it ran from Athens to Augusta on the Savannah River and served the cotton plantations. The cotton bales could be loaded on the riverboats at Augusta and floated down the river to the port at Savannah. By 1837 there were at least two more railroad charters granted. The first charters carried the stipulation that no cars would run on the Sabbath. The railroads did little to improve the declining economy of Lumpkin County. (6)

The city of Milledgeville had been established as the Georgia capital in 1803. The location seemed appropriate at the time as the Oconee River is 550 feet wide at this location and provided a safe dock for boats. It was in the heart of cotton country. "The steamer with two tow boats can bring about 150 tons at a trip." (7)

By 1850 many new railroad tracks crossed the state. The land confiscated from the Cherokee Nation had for the most part been taken up by cotton planters. Transporting bales of cotton by rail was more efficient and less expensive than river travel. Slaves could carry loads of cotton to the railroad siding tracks without leaving plantation grounds. At first only two, but soon many rail lines converged at the round house in Atlanta. The town would be named the capital city of the state in 1867. (8)

While gold mining intrigued the attention of many, a financial crisis known as the Panic of 1857 swept the entire United States. Banks and financial institutions closed their doors. Railroads consolidated and many branch lines were closed. The general population lost faith in the "paper" money printed by the government and demanded "hard cash" for any goods or services rendered.

The problems of slavery loomed ever larger in Georgia affairs. Northern abolitionists grew in number. So many petitions and letters were directed to the House of Representatives in Washington that in the 1830s the House established a "gag rule." This decision meant no letter or petition concerning a single slave or many slaves could be read, and no discussions about slavery would be permitted. The Senate on the other hand was the scene of great debates and political manipulations so that every new "free" state admitted to the Union would be balanced by a "slave" state. James Buchanan was elected President, taking office in 1857, but his policy of avoiding controversy and the Missouri Compromise he favored led to confusion and violence. The policy was an attempt to achieve balance, but Senators had to decide what to do with free black men, how to treat the recapture of a black man from a slave state but who now lived in a free state, and whether any slave could be counted in the census or would simply be listed as property.

A "country lawyer," Abraham Lincoln from Illinois became known as a fiery speaker. People flocked to see and hear this tall man who stood 6'4" tall. His high-pitched voice carried clearly to the ears of a large audience. Men and women stood for a long time, usually in an outdoor space, eager to hear him. He ran for office against Stephen A. Douglas and although Douglas won the contested seat in Congress, Lincoln's words were the ones that voters remembered. His theme, repeated over and over, was the inhumanity of slavery and the schism it was creating in the nation.

As the clouds of war intensified, stories of gold in the Rockies seemed ever more alluring to the men of Georgia. Eight years earlier prospective gold miners, seeking to join a wagon train, had spent a minimum of $200 to equip themselves for a journey across the continent. In 1858 the Rocky Mountains were little more than half the distance of a California trip, but preparing for this journey was still a major investment. Elizabeth Ralston wrote to her cousin, Emily (Duncan) Beck in Indian Territory about Lewis' desire to "go prospecting." Emily's husband, Baptist minister John Beck, made plans to recruit a group from the Nation. A total of 16 men assembled in Auraria, Georgia. The Russell brothers, Green, John, Oliver and Levi were present, along with Lewis Ralston. Green, by common consent, was named captain of the group.

The group started west. After his return from California, Green had purchased a farm in Kansas Territory that he named Rock Creek. It was near the small village that would be named Leavenworth. During the winter of 1857, two of the Russell cousins, Jim and Robert Pierce, had lived on the farm and they would join the wagon train. The new McNair prospecting party reached Rock Creek farm and waited for Cherokees from Tahlequah to join them. Two other men, T. C. Dickson and Luke Tierney, heard of the proposed trip, and they walked to the farm and asked to join the group. Both later wrote of their adventures and Luke Tierney's account was published in 1859.

We can surmise that an impatient Lewis was eager to push on to Ralston Creek. He was not as well provisioned as the Russells who had California gold to spend, and Elizabeth was expecting a baby who would be a boy named James. He had suffered only disappointment from the Georgia and California gold strikes. This trip would be different – this time, he was sure a fortune in gold would be his.

Ralston Creek

Clear Creek Canyon

Chapter VIII

❧ Return to Ralston Creek ❧
1858

1870 location Arvada, Colorado

The lack of recognition for the site of Ralston's discovery is due in part to misinformation from the very beginning. The John Lowery Brown diary is the best original source for events in 1850. However, the existence of the diary was unknown for many years.

A letter sent to the author explains that John Lowery Brown had returned from California and the small booklet was a family treasure but was not known to historians. Elizabeth B. Freeman, a great granddaughter of Brown, told this story to Catherine B. Gist.

> Spi Trent, grandson of John L. Brown, put the historic diary in a secret drawer in a roll-top desk. Later, when he moved from the office, he left the desk and the diary, behind.

> In 1924 Mrs. Clover Barrowman, also related to John L. Brown, was riding to Spavinaw and overheard two strangers discussing Cherokee history. One man told the other about a diary he had found in an old roll-top desk. Clover interrupted their conversation and described the Brown diary which she had seen many years before. The precious document was returned to the descendents of John L. Brown.

> Catherine Gist, Mrs. E.W., granddaughter of John Lowery Brown inherited the diary and she donated it to the Thomas Gilcrease Institute of American History and Art in Tulsa, Oklahoma. The author has held this important journal in her hands.

In 1924 Muriel H. Wright researched the entries, wrote historical background and prepared a typescript. All entries are reproduced exactly as they appear in the original diary. Her work was published ten years later in 1934 by the Oklahoma Historical Society, and a copy was given to the Colorado Historical Society.

One of the best accounts of the 1858 Ralston/Russell return party is the diary written by Luke Tierney. He wrote a detailed account of his journey and it was published in 1859 by D. C. Oakes and S. W. Smith. The Tierney account was republished in 1941, edited and annotated by LeRoy R. Hafen.

James H. Pierce wrote a "paper" in 1885 that was published in THE TRAIL in 1921. A third document is that of Colonel T. C. Dickson who related his memories of the 1858 gold prospecting trip to J. D. Miller in 1911 and it too was published in *THE TRAIL*. Credit is given to the named authors in Sources. [1]

Pierce: Men who left Georgia with Green Russell on April 5, 1858:

W. G. Russell - Green	R. J. Pierce - cousin
J. O. Russell - Oliver	Samuel Bates
L. J. Russell - Levi	Solomon Rowe
James Pierce - cousin	"six or eight more"

Lewis Ralston was included in the "six or eight" more. These individuals were probably Cherokee and deemed not worthy of being recognized by name. Lewis Ralston's wife was Cherokee – he was a white man.

Jerome Smiley, who interviewed Dr. Levi Russell, said the group included: [2]

W. G. Russell	Joseph McAfee
J. O. Russell	Solomon Roe
L. J. Russell,	Samuel Bates
Lewis Ralston	John Hampton
William Anderson	

Dickson: (Leavenworth, Kansas Territory, 1858) ". . . I came across a man named Green Russell . . . who was fitting up teams and wagons and getting supplies preparatory to going out to the Rocky Mountains to search for gold. . . he expected to start from the 12th to the 15th of May. He (Russell) said that on his trip to California in 1849 a man by the name of Ralston . . . in his party . . . found some gold."

Some of the above is correct. Ralston did find gold. However, Ralston found gold in 1850, not 1849, and Ralston was not a member of Green

Russell's 1849 prospecting party. Green was not on Ralston Creek in 1850 – Lewis Ralston was.

Pierce: (part of the original group from Auraria, Georgia 1858) "Beck, in company with Louis (Lewis) Ralston in crossing the plains in the year 1849, going to California had found gold on Cherry Creek and also on Ralston Creek in small quantities." . . . Russell had found gold on the North Platte and on the Sweetwater while . . . going to California."

The finding of gold in Ralston Creek was June 22, 1850, not 1849. Russell was not present. The North Platte and Sweetwater rivers were streams crossed by the Oregon Trail in Wyoming.

Green Russell, may have been trying to establish a connection with the 1850 discovery on Ralston Creek. Russell was not part of the McNair/Taylor wagon train. John Beck, a Cherokee pastor, and Lewis Ralston were.

The authors of *Cherokee Trail Diaries,* a massive annotated historical work, spent years researching the 1849-1850 prospecting parties that left from the Oklahoma area and traveled the Cherokee Trail. They found no record of a Russell party in 1849 or 1850. There are only two listings for Russell in the 418 pages of the Fletcher's book. They found mention of Captain William H. Russell, (Green's name was William Greeneberry) from Missouri who guided a prospecting party up the Platte River Trail in 1849. They quote from the John L. Brown diary listing a Russell (no first name) who died from cholera August 7, 1850 as the McNair/Taylor party traveled through Utah. (3)

Tierney: Luke Tierney joined the Lewis Ralston/Green Russell party at Rock Creek, Kansas Territory, on May 16. 1858. His account is one of the most detailed and accurate gold trail diaries that have survived. The Ralston/Russell wagon train was camped at the "junction of the Smoky Hill and Republican Forks of the Kansas River" when Tierney and two others joined the group. The total party then consisted of "20 men, 4 wagons, 10 yoke of cattle and 3 horses."

Late in May the Ralston/Russell wagon train started west. Some of their horses ran away, and although members of the party searched for several days the ponies were not recovered. On June 1 they reached Pawnee Fork (near present-day Garfield, Kansas). Here they found a "village" of Cheyenne Indian tents. The friendly natives traded with the gold miners. On June 3 they were joined by the group headed by Cherokee pastor Reverend John Beck and Cherokee attorney George Hicks that had been organized at Tahlaquah, Indian Territory. The two wagon trains met . . . "where the old Santa Fe Trail . . . crossed the Arkansas River." The combined assembly – " consisted of 70 men, 14 wagons, 33 yoke of cattle, 2 horse teams and about 20 ponies."

On June 12, following the north bank of the South Platte River, they passed near the new Bent's Fort at Big Timbers and they could see snow-topped mountains in the distance. Tierney said the "weather was very hot." They were following the trail pioneered by Captain Louis Evans in 1849, rumbling along in the rutted tracks of the McNair/Taylor wagon train of 1850. On the "Sabbath, 20th of June," they covered twelve miles in the morning to reach Black Squirrel Creek. They stopped for the afternoon of rest. Here they filled the "water vessels" and killed six antelope, a welcome addition to the food supply.

They reached the headwaters of Cherry Creek on June 22, 1858. Where the trail crossed the creek "one of our men taking a pan full of gravel from the bed (of the stream) washed it and found several particles of gold."

Tierney certainly knew Green Russell. The man finding the gold is unnamed. Tierney would have used his name if it had been Green Russell. Later, according to historian LeRoy R. Hafen, a gold camp named Russellville was established at the site in 1859. [4]

On this day, June 22, they were nearing the confluence of Cherry Creek and the Platte River. None of the diary keepers or train members mentioned that this was the date of Lewis Ralston's discovery eight years before. On the 23rd Tierney wrote they had "at noon . . . a sumptuous feast of various kinds of wild game." They studied the roaring floodwaters of the Platte where "the depth of water rendered it unsafe to attempt a crossing."

"The day following (June 24) . . . by wading, swimming and sounding, struck off (found) the safest and only fording place where we could effect a landing on the opposite shore." Eight years before, in 1850, the spring flood had been at its height when the McNair wagon train reached the confluence of Cherry Creek and the Platte. The 1850 party had traveled east along the Platte River banks to find a low spot.

The 1858 Ralston/Russell train found a crossing nearer the South Platte River/Cherry Creek confluence.

Tierney: June 24, 6 P.M. (1858)

"We continued our march over high rolling prairie and down some most stupendous banks, a distance of about five miles, at the end of which we came to a large creek, known as Long's creek, (Vasquez/Clear Creek) where we found great difficulty in effecting a crossing. The water was about five feet deep, very wide, and the current unusually swift. However we crossed over in safety and continued our march, reaching our destination –

RALLSTON'S CREEK

About 6 o'clock P.M. Here, according to the statements of the returned Californians we were in the immediate vicinity of the gold mines." (Tierney used capital letters to convey the importance of finally reaching their destination. This is his spelling of Ralston.)

On June 25 Tierney listed the group: "104 men from different sections. Georgia 10, Missouri 27, Cherokee Nation 58."

Dickson: June 20

"We experienced a great difficulty in crossing." (South Platte River). We proceeded on to Ralston Creek."

Pierce: May 23

"We . . . camped at the mouth of Cherry Creek. . . . We then crossed the Platte, went to Ralston Creek and camped there where Beck and Ralston had found gold in the year of 1849." (Should be 1850)

Historians rejoice that John Lowery Brown was so very specific about dates in his diary of the 1850 trip.

Lewis Ralston must have been greatly pleased to return to his creek, but none of the writers solicited his comments. By this date, they had been joined by men from the "Missouri Company" and the group now included 104 men, all focused on one thing – the acquisition of gold. The men fanned out across the gulches of Ralston and Clear Creeks, but quickly discovered that gold panning was yielding meager results. After days of effort the men on the average were collecting only about 25 cents each per day in fine "sifted gold."

Pierce: "We thought if we could make some sluice boxes we could make it   pay. We took the bottom boards of some of our wagons and made two of these. We then built a dam across the stream, (Ralston Creek) took out a ditch and commenced to sluice the gold-bearing dirt."

Some of the miners hiked further west on Ralston Creek or tested other ravines. Green Russell, Luke Tierney and Jim Pierce decided to prospect up the rock-lined walls of Clear Creek.

Pierce: "We tried to make our way up Clear Creek, the party on foot in advance while I was trying to climb along the side of the mountain with my pack animals. The men on foot had not gone over a mile when they called down to me to turn back as they could go no further, saying –'a bird could not fly up that canyon.'"

They would later learn that the mountains surrounding the waters of Clear Creek held a vast golden treasure, but this day they could do no more.

In the last days of June the men began to reassemble at the Ralston Creek camp. Some of the miners had explored the banks of the Boulder, Big Thompson and Bear Creeks. No one had succeeded in making a strike, everyone was discouraged.

Tierney: July 3, 1858
"On the evening of the third of July . . . the company all assembled together for consultation . . . all were silent. . . considering whether to return home without further search or remain and risk further disappointment . . . Sunday, July 4, the greater part of our company . . . were making active preparations to return to their homes."

Ralston gold discovery!

June 22, 1850

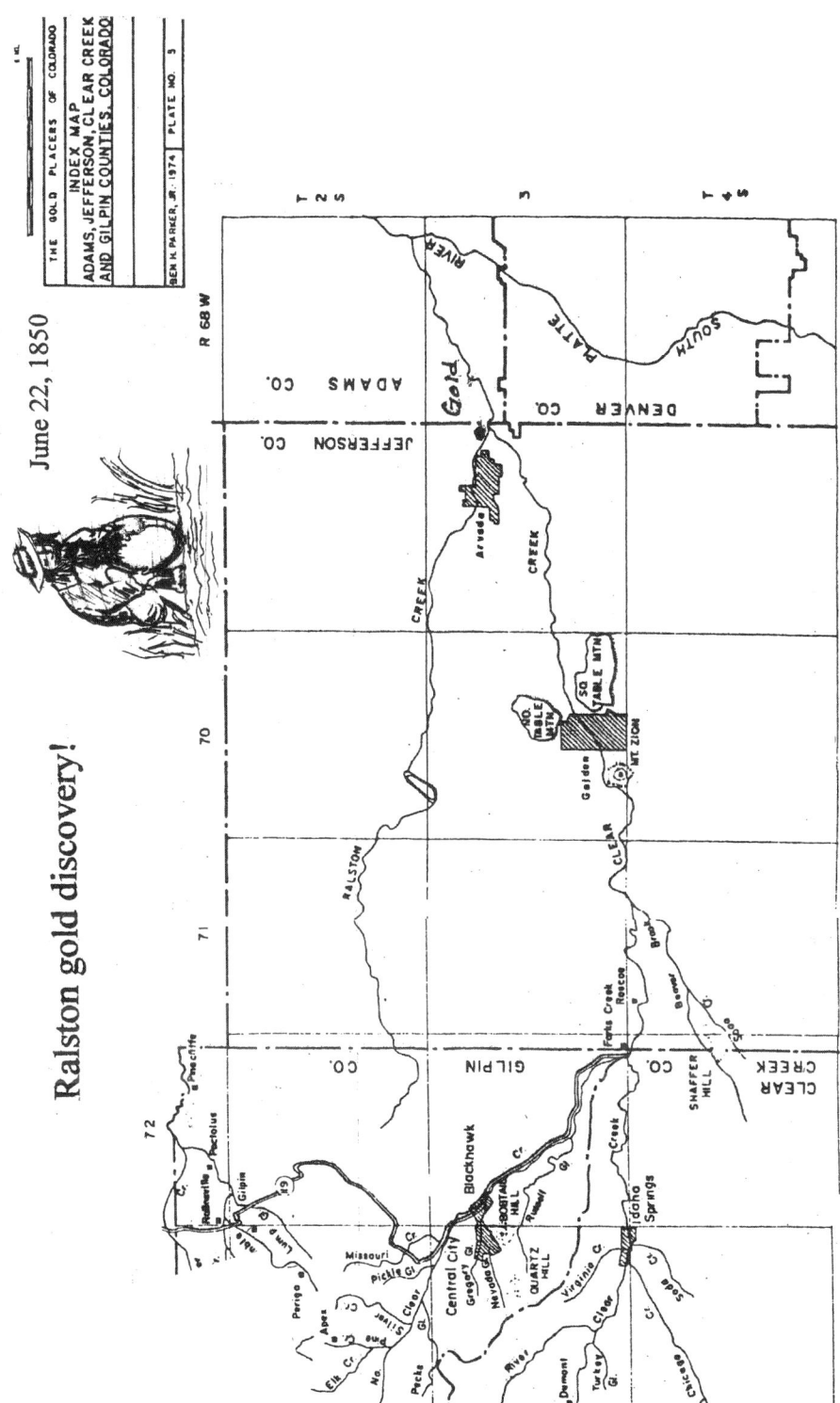

Map: Ben K. Parker Jr.

73

On July 5, "10 o'clock," all the Cherokee and Missouri companies "took their departure." Remnants of the once proud group of 104 men decided to return to the Platte River. The supply wagon was capsized as they attempted a crossing.

Pierce: . . ."we came near losing the wagon and all our money and summer provisions . . .We managed to get most of our rations out but our rations were damaged considerably by the water."

Losing a great part of their food, the flour sacks dripping water, most of the remaining members of the Green Russell party announced their intentions to leave that day for home.

Pierce: "Green Russell, though a man of iron nerves, was shocked to see his cherished scheme about to become a failure. . . he said, 'Gentlemen you can all go but I will stay if but two men will stay with me." Of the group, only 12 men stepped forward, including his brothers and cousins, "and all the balance went home."

Dickson listed the names of the thirteen strong men who were determined to unlock the secret of Rocky Mountain gold:

From Georgia:
W. G. Russell   V. W. Young, Iowa
J. O Russell   Theodore Herring, Kansas Territory
L. J. Russell   William McFadding, N. Carolina
J. H. Pierce   William Remnins, Pennsylvania
R. J. Pierce   J. T. Masterson, Kansas
Samuel Bates   Luke Tierney, Kansas
Solomon Roe

J. H. Pierce later listed Valorious Young and not Remnins. William McKimons said he was part of the group of 13.

Lewis Ralston was one of the men who turned toward home. He may have turned back with the first group called the "Cherokees." He certainly turned back after the disaster with the supply wagon. He probably was out of food and terribly discouraged. Lewis Ralston played no further role in Colorado's history.

That same day the thirteen men said farewell to their comrades and Tierney wrote: "marched up the Platte (west) in the direction of the mountains. . . . on the very day the last party deserted us . . . we found the first cheering prospect of gold – the diggings yielding about $10 a day for each hand." This was fine "drifted gold." Ten days later they found another small pocket that panned out $12 to $18 for each man, and finally a third "deposit" which yielded about the same as the first two. This third strike was worked for 10 days.

Historian LeRoy Hafen, published in 1948, named the third site Placer Camp and near this camp some gold seekers later founded Montana City, one of the temporary tent camps that had a short life. Hafen located Placer Camp on Dry Gulch/Little Dry Creek in present-day Englewood. (5)

The thirteen prospectors eagerly worked the new "mines" but the gold was "drift gold" or "flour gold" and these experienced miners knew it came from a rich vein somewhere upstream. The gold they were panning had been released from rock by erosion and washed down by action of the mountain streams.

They were in the middle of a beautiful wilderness. They had found gold and there were no other human beings near to learn their secret. But this isolation would last for only a few more days.

"On the last day of July" a group of horsemen rode up to their camp. Among them was John Cantrell from Westport Missouri. He had been at Fort Laramie (Wyoming) selling whiskey and supplies and heard that some white men were panning gold along the Platte River. He and his companions rode south to investigate.

Some stories say Cantrell dug out some "paying dirt" with his hunting axe and panned gold with his frying pan. The group soon went on their way home to Missouri, but Cantrell contacted a newspaper editor. On August 26 the headline of the *Kansas City Journal of Commerce* shouted: THE NEW ELDORADO! GOLD IN KANSAS TERRITORY! (6)

Back on Little Dry Creek, Green Russell knew they would not be all alone much longer. His crew was suffering from hard labor in the blazing sun and had very poor rations to sustain them. Nevertheless they formed small groups and traveled a wide area looking for the Mother Lode that had birthed their gold dust.

Some traveled north, even into lands that would become Wyoming, where they met some government troops and shared a meal of bear meat. Finally they returned to the confluence of Cherry Creek and the Platte River on September 20, 1858.

Auraria and Denver City

The beginnings of Auraria and Denver City have been told in detail by other authors. These writers, trace, in part, the efforts of the Green Russell party in June of 1858 and the founding of Auraria only five months later. There are few who wonder why the experienced Russell miners were inspired to come to the wilderness in the first place. There are few mentions of the first discovery by Lewis Ralston in 1850.

On September 20, 1858, when Green and his men returned to their rendezvous site at Cherry Creek and the Platte River, they were amazed to find new gold seekers lured to the spot by the Kansas City publicity. There were also the colorful tents of Arapaho Indians clustered on the creek banks. The group of 13 made fateful decisions. They had $500 in gold dust and Green and brother Oliver, along with cousin Jim Pierce, would ride back to Georgia to recruit more men and purchase supplies.

Levi Russell, Luke Tierney, William McFadding and William McKimmons would ride south to try to buy some food. Fort Garland (originally Fort Massachusetts) had been newly commissioned and relocated. Russell probably learned about Fort Garland from the government troops in Wyoming. (7)

Before Green and the others left for Georgia the idea of founding a new town was discussed. They would have to survey and claim the land, and a structure had to be built to hold the claim. Green and Levi, with their companions, rode off in separate directions. The remaining five men, along with assistance from just arrived newcomers to the area, built a sturdy double log cabin within the point of land where Cherry Creek joined the Platte River. They selected a spot where the shelter of the two rivers would alert the cabin dwellers of approaching danger.

Dr. Levi and his party rode to the new (1858) Fort Garland and Levi traded his gold watch for flour and beans. Green and his group moved as quickly as possible to get their gold home to the mint in Dahlonega, Georgia. When Levi and his group returned to Cherry Creek the stalwart log house was finished. Levi later said, "It was the only building in sight."

The prospectors began town planning. On November 1, 1858 they wrote up the necessary charter for a new Town of Auraria. Levi suggested the name of Auraria in honor of the first "gold" town in Georgia. They made a list of stockholders. The list includes 100 names, but many of these signatures were added at a later date. The charter was filed with the territorial government in Topeka, Kansas Territory, but only the Denver records have survived. (8)

Auraria was the first "gold" town in what would become the state of Colorado. It was not the first town in the state. San Luis, founded in 1852 by shepherds from New Mexico, claims that title.

Later the same month, November 16, a group from Lawrence, Kansas Territory, rode up. They had seen nuggets displayed by an Indian named Fall Leaf and had organized a prospecting party. They carried documents and permission to colonize from the Territorial Governor of Kansas Territory, James W. Denver. The governor had established the boundaries for a new county in Kansas Territory to be named Arapahoe and listed the men who would be the legal officers.

The Lawrence men were shocked to see the sturdy log house and tents of the town of Auraria already existing on the west bank of Cherry Creek. The Lawrence prospectors disregarded the claim of another group that had erected tents for a settlement named St. Charles on the east bank of Cherry Creek. The St. Charles group had not erected a building to hold their claim. The St. Charles site was renamed Denver City in honor of Governor Denver, and the legal charter was completed on November 22, 1858. Out west on Clear Creek, where the foothills raised toward the towering mountains, Arapahoe City and Golden City were established 7 months later in 1859. (9)

The town leaders of Auraria and Denver City were in constant turmoil regarding civic affairs. The two small villages, probably not more than 50 citizens each, could not agree on leadership or control. The population of both settlements was in constant flux as new gold seekers arrived and then left immediately for gold "diggings" in the mountains. On April 5, 1860 an election was held, with Denver City emerging as the most popular name for a unified town. A torch light celebration was held at midnight in the center of the Cherry Creek bridge.

Ambitious town leaders met in Richens Wooten's Auraria store on

October 24, 1859, and voted to send Hiram J. Graham to Washington D. C. He was commissioned to ask for the creation of a new State of Jefferson. They also voted to send A. J. Smith, as a representative to the Kansas Territorial Legislature, and he was to ask for the creation of the Territory of Jefferson. By simultaneously seeking both statehood and territorial recognition they hoped to achieve one of the two. The U. S. Congress did not favor the name of Jefferson and the important issues of "slave or "free" state filled every agenda. Finally the federal Congress created the State of Kansas on January 29, and one month later, February 28, 1861, they carved out the Territory of Colorado, without any reference to slavery. The Civil War began two months later on April 14.

The great gold rush was truly born in 1859 with the announcement of these strikes:

June, 1858	Arvada – sluice mining
	Ralston and Clear Creeks
	Lewis Ralston
1859	
January 7	Chicago Creek (Idaho Springs)
	George A. Jackson
January 16	Gold Hill (Boulder)
	Captain Thomas Akins
May	John H. Gregory Blackhawk and Central City
June	Russell Gulch
	W. G. Russell (Green)

Other important gold finds were:

1859	Golden
1859	Breckenridge
1891	Cripple Creek

There are few truly accurate accounts of the gold found in the Clear Creek area in the early days. Miners carried their gold in a small leather pouch, in a goose quill, in special pocketed vests, or any small safe container. Some early merchants allowed customers to deposit gold in the store vault or safe. Much of the gold dug by the early miners was carried back to their homes in the East. A branch of the United States Mint was established in Denver

City in 1862, but the present building was not constructed until 1895. The Colorado Bureau of Mines was established the same year.

In 1926 Charles W. Henderson's paper, *Mining in Colorado* was published by the United States Department of the Interior. Henderson used a variety of sources and developed a chart which includes the years 1859-1865 and indicates that placer mining in Clear Creek County totaled $2 million dollars for these six years. Much of the gold mined in the early years was simply not reported. Some historians have estimated that the total Colorado output during this time was closer to $3 million dollars. [10]

The great 1859 gold rush mandated the creation of government. Many people came, not as miners, but as suppliers for the needs and wants of the miners. These early merchants brought their families and with them and the necessity for rules, laws and schools.

June 22, 1850	Lewis Ralston-gold discovery
June 24, 1858	Return to Ralston Creek
November 1, 1858	Auraria founded
November 22, 1858	Denver City founded
June 9, 1859	Golden City founded
February 28, 1861	Colorado Territory
June 8, 1864	Central City founded
December 1, 1870	Arvada founded
August 1, 1876	Colorado Statehood

In 1870, twenty years after the discovery on Ralston Creek, the Colorado Central Railroad was built linking Denver City and Golden City. The great boon of railroad transportation sparked the founding of Arvada by Benjamin Franklin Wadsworth on December 1, 1870.

Lewis Ralston was not part of these events. He may not have known about the John Lowery Brown diary. Through the years conflicting accounts have been written about the first gold discovery on Ralston Creek. Medical doctor Levi Russell, living in Texas, was interviewed by LeRoy Hafen and his remarks were published in 1946. Hafen quoted Levi who said that the Russell brothers had not known of the gold find on Ralston Creek until they returned to Georgia from their second California trip in 1857. [11]

Ralston's wife, Elizabeth, and her cousin Emily Beck, certainly knew who discovered gold in 1850. No early Colorado historian ever questioned them. Hollister said that Green Russell learned of the gold from Lewis Ralston's brother, Samuel. Lewis may have had a brother-in-law Samuel Simons, a name listed by diary-keeper Brown, but there is no other mention of Simons. Lewis did have a brother Samuel, born in 1823. As an adult, his brother Samuel resided in Gilmer County, Georgia. There are no records to indicate that he took part in any gold mining efforts. (12)

Lewis Ralston's farm in Lumpkin County and Green Russell's plantation at the Hightower Crossing on the Etowah River were less than 15 miles apart. Green Russell probably learned of the Ralston Creek gold from Lewis Ralston himself.

In 1850 the wonderful land that would become the state of Colorado was a treasure chest just waiting to be unlocked. Ralston's gold discovery on June 22, 1850, opened the door to Colorado's destiny and brought the gold prospectors of 1858. These "58ers" initiated the gold rush of 1859. The prospectors in turn brought the wagon road and railroad builders, the merchants, farmers, civic leaders, teachers and settlers that together crafted a magnificent state.

CHAPTER IX

✎ PRIVATE LEWIS RALSTON ✎

At some point after the "lay bye" on July 4, 1858, Lewis Ralston began the long journey back from the untamed wilderness that would, in 1861, be named Colorado Territory. He was deeply discouraged by the results of his second gold prospecting adventure. His spirit lifted and his steps quickened at his first glimpse of the gently rolling Blue Ridge Mountains. He could forget his failure to find a golden treasure, forget the terrible majesty of the Rocky Mountains in the West.

Elizabeth and the children welcomed him back to Chestatee Old Town. He may still have owned parts of his estate that in 1836 included four separate plats of land in Lumpkin County and another small holding in Forsyth County. He had been "awarded" damages of almost $6,000 when the federal government seized Cherokee lands but there is no record to show that he received any of this amount. [1]

No one marked the day of his homecoming. The local newspaper, *The Dahlonega Signal,* July 6, 1858, lists the names of those called to serve on the grand jury for one week. More than 20 men are listed and Lewis Ralston is the second name on the list. He could not have served because Lewis was still camped on Ralston Creek on July 4. [2]

The Ralston's son, Lewis Jr. was 21 years old in 1858. He was the seventh child and the third son in the family. John, the first son, named for Ralston Sr.'s father, John Tate Ralston, was 29 and still living on the farm. Alexander, the second son and fourth child, died in infancy. He was named for his grandfather, Elizabeth's father, Alexander Kell, who was born in Scotland. Scotland was also the ancestral home for Lewis Sr.'s father, John Tate Ralston. Alexander Kell's wife was related to Cherokee Chief Benge, as was Elizabeth Ralston. Lewis Ralston Sr. left his sons, John and Lewis Jr. to manage the farm while he prospected in the West. [3]

The Ralston family and their neighbors were well aware of the Abolitionist movement developing in the Northern states. The efforts of those who would abolish slavery were strengthened by the book written by Harriet

Beecher Stowe. *Uncle Tom's Cabin* was created by Mrs. Stowe sitting on her kitchen steps after her daily housework was done. The book was published in 1851 and it became the rallying voice for the anti-slavery cause. The book was banned in Southern states, but news of the anti-slavery movement spread by word-of-mouth to Georgia's plantations.

When his father returned home, Lewis Jr. was 21 years of age. He had fulfilled his duties to his father and mother. He married Eliza Postell, a non-Cherokee woman. They were married in Lumpkin County, Georgia, on December 20, 1859. Soon after their marriage the young couple started west. Lewis Sr. and Elizabeth continued the never-ending farm work with the help of the remaining children – six girls ages 5-27 and three boys ages 2-14.

Lumpkin County residents were shaken and angry at reports detailing the furious debates about slavery in the federal Congress. Abraham Lincoln, known as "Honest Abe," representing the Whig party, had been elected from Illinois to the House of Representatives in 1846. Although he did not seek re-election he was asked to engage in a debate with Senator Stephen A. Douglas. Douglas promoted repeal of the Missouri Compromise, the 1820 law limiting the spread of slavery. Douglas spoke first for two hours. Lincoln's much shorter speech held words that everyone remembered when he called for an end to the "monstrous injustice of slavery." Lincoln felt he had failed when the Kansas-Nebraska Act became law in May of 1854. This law allowed potential new states to choose themselves whether or not to be "free" or "slave." (4)

Lewis Ralston and most of Georgia rejoiced when Lincoln, representing the Whig party, failed in his first attempt to win the election for the Senate seat from Illinois. In 1855, Lincoln was instrumental in the founding of the Illinois Republican party, and in 1858 he was again candidate for the Senate. He gave a speech which included the famous quotation: "A house divided against itself cannot stand. . . the government cannot endure permanently half slave and half free." Large crowds gathered to hear him, but in the 1859 election, Douglas retained the Senate seat, defeating Lincoln 54 votes to 46. (5)

The wealthy plantation owners of Georgia tried to keep Lincoln's message from reaching their slaves. However, the powerful idea of freedom could not be stifled and spread from plantation to plantation by drum beat and furtive conversations away from the "big house." The voters of Lumpkin

County furiously rejected Lincoln's words. Even those citizens who did not own slaves, and by this time this was probably true of the Ralston's, the South supported slavery. Men of the South claimed a "states right" to own slaves. Georgia's economic well-being rested on the backs of slaves who dug, planted, weeded and harvested cotton - the South's major cash crop. Slavery had been an institution in the South since 1619 when John Rolfe, husband to famed Indian princess Pocahontas, brought in the first boatload of black men from Africa.

A book that greatly increased anger in the South was *Journal of a Residence on a Georgia Plantation* written by Fanny Kemble. The book was not published until 1863, two years into the Civil War, but it strengthened the positions of both armies. Fanny wrote with the authority of an eyewitness. Catherine Clinton's meticulous research in her book *Fanny Kemble's Civil Wars* validates Kemble's work.

Kemble was an acclaimed British actress. In 1832 she sailed to the United States for an American tour. Her beauty and fame attracted many suitors. Most ardent and persistent of them all was Pierce Butler. He followed her from city to city and they were married June 7, 1834. Pierce was part owner of the Butler Plantation, which included property that stretched along the Altamaha River near the port city of Darien, Georgia. His family also owned a stately mansion centered proudly on 300 acres in Branchtown, a few miles from Philadelphia.

The Butlers owned more than 1000 slaves all confined to their various plantations. Butlers' property included most of St. Simon Island, and all of Butler Island and Little St. Simon Island, as well as a large property called Experiment near Hampton Point – all located in Georgia. The slaves produced Sea Island cotton valued for its strong fine fibers, and along marshy beaches harvested a superior Sea Island rice.

Fanny knew almost nothing about slaves and plantations before she married Pierce Butler. Great Britain had abolished slavery throughout all colonies and territories in 1807. To give voice to her shock and revulsion, in 1835, a year after her wedding, she wrote *A Treatise Against Negro Slavery*. The paper had very little public circulation. In the following years, after the birth of two daughters, and with increasing estrangement from Pierce, she tried to secure a divorce. The national laws of the 1800s prohibited divorce without the consent of the husband. The divorce was finally granted. Fanny had to pay Pierce a large sum of money and give up all contact with her girls (as adults they sailed to England and lived with her).

Fanny booked passage home to England in 1845 and resumed her acting career. She had to resort to court action to retain her income because, in spite of the divorce, Butler, by law, had control of all her finances. She finished her book *Journal of a Residence,* but Butler, protected by existing laws, blocked her efforts at publication. The book was finally published by both British and American printers and was distributed in 1863. [6]

The book was a literary sensation in the Northern states and rejected as false by readers in the South. Many historians today feel that her descriptions were a true assessment of the lives of slaves in Georgia. Soldiers in the Confederate army were energized to greater efforts to prove Fanny Kemble wrong.

Abraham Lincoln was elected President of the United States of America in 1860 – President of all 33 states of the Union. Lincoln and his vice-president, Hannibal Hamlin, received 180 of the possible 300 Electoral College votes. They did not receive a single vote from the eleven Southern states. He said, "If you stand by me I will stand by the Union."

Lincoln and his supporters had hoped to hold the nation together, but the South had heard enough. Only a month after the election, before Lincoln could be inaugurated in March, on December 20, 1860, South Carolina withdrew from the Union. On January 19, 1861, Georgia followed South Carolina, as did Florida, Mississippi and Alabama. Jefferson Davis was installed as President of the Confederate States of America on February 18, 1861. Two months later, on April 14, 1861, Fort Sumter, a national military post, was taken by Southern militia. The fort, on an island in Charleston Bay, was only 60 miles north of the Butler Plantation at Darien, Georgia. The Civil War had begun.

The Ralston's and Civil War

During these troubled years John Tate, the oldest son of Lewis and Elizabeth left the farm. There are few records of his later years. He is mentioned as residing in Indian Territory (Oklahoma) in an account written many years later. Lewis Jr., who married in 1859, was farming in Missouri.

On June 7, 1860, the census taker reached Lewis Sr. and his family in Lumpkin County. Lewis, a farmer, gave his age as 56 that year and Elizabeth's age is written as 59 (she was really 48). Their children were listed as Frances, Nancy, Elizabeth, Henry, Elizar, Zachariah, James, Josephine and Robert; ages 27 to only 4 years old. The value of Ralston's real estate was $1,500, and his personal estate was $200.

The Ralston family, as did many others, kept to the exacting rotation of raising grain crops, feeding their cattle, cultivating a garden and maintaining their orchard. They could still regularly drive their produce to market in Atlanta in spite of the gathering war clouds. This uneasy peace existed only until South Carolina left the Union in 1860, followed shortly by other Southern states including Georgia. Dahlonega's young men quickly joined the Confederate army, but during the first years of the war, the pastures and cotton fields of Georgia were spared the devastation ravaging the cultivated fields in the North.

Atlanta, approximately 50 miles from Dahlonega, was a bustling transportation center by 1860. In 1813 it had been the site of an U.S Army post named Standing Peachtree. A small village grew which was named Terminus because of the new rail traffic. The first two railroads, the Central Railroad and the Georgia Railroad converged at the site in 1837. In 1843 Terminus was renamed Marthasville, and as rail transportation increased, the much larger town was re-titled Atlanta in 1845. In 1853 it was named the county seat of Fulton County and Atlanta replaced Milledgeville as the state capital in 1868. [7]

Lewis Ralston and the farmers of Lumpkin County had been greatly encouraged by the building of Atlanta's rail center. They had a larger market for the products of their farms. The misty valleys of the Blue Ridge Mountains, rising slowly upward from Lumpkin County's foothills, were not conducive to raising cotton. Conditions were ideal for growing peaches. Georgia farmers knew their peaches had a unique flavor and appearance and the fruit commanded high prices in Northern markets. The Ralstons carefully tended their orchard.

The Confederate capital city was established at Richmond, Virginia in 1861. As the battles of the Civil War developed, Atlanta, Georgia became a major supply depot for the Confederacy.

Dahlonega's young men responded eagerly to war's beginning on

April 4, 1861. From *The Dahlonega Sentinel* newspaper for April 5, 1861: "The Dahlonega Volunteers commanded by Captain Harris left . . . on April first, enroute for Macon (Georgia), where they are to be mustered into service." The company and a large group of citizens met at the courthouse for prayers and an "eloquent speech." They then "formed in a line" carrying a silk Confederate flag made by the "young ladies" of the town. They received a "parting salute from the anvils." The number of these first volunteers is not known. By July, three months later, 50 more Dahlonega men had volunteered. They, too, were given a rousing farewell with a flag and a "stirring speech." (8)

The Ralston's son, Henry, age 14, was the oldest boy still at home. Henry was needed to help with the farm work and father Lewis probably discouraged any desire Henry might have had to enlist. The two oldest sons, John and Lewis Jr. had already moved westward, moves that may have been triggered by pre-war tensions.

An interesting item was recorded by historian Andrew Cain: He quoted the Dahlonega newspaper regarding the clothing made by the Ladies Aid Society. A Lumpkin County woman, "Mrs. Garvin" made a complete outfit, 20 pieces in all, including "socks," for a soldier in the field and "better cloth never came from the hands of a patriot." Mrs. Garvin had spun the fabric herself. The Cherokee women since early days had learned to use spinning wheels furnished by the first missionaries. *The Dahlonega Sentinel* wrote of thread from cotton spun by a wool factory in Amicalola, which "can answer all the purposes of domestic use." The 1835 census had indicated that mother Elizabeth Ralston was a weaver. The oldest daughters of Lewis and Elizabeth, still living at home, were Frances, Nancy, Elizabeth and Agnes. They were 19 to 30 years of age and Elizabeth may have taught them to spin and weave. (9)

Some of the major Civil War events that brought war closer to Georgia:

1859

October 16   Harper's Ferry

1861

April 14   Fort Sumter

July 21   Bull Run/Manassas, Virginia, Confederate victory

July   Blockade of all Southern ports by the Union

1862

March 9	Battle of the first two iron-clad boats built in the nation. Merrimac (Confederate) Monitor (Union) Hampton Roads (Newport, Virginia) Both sides claimed victory
April 6-7	Shiloh, Union victory
June 25	Union attempt on Richmond failed, Confederate victory
August 9	Union second attempt on Richmond failed, Confederate victory
Sept. 17	Antietam, Union victory
Dec. 13	Fredericksburg, Confederate victory

1863

January 1	Lincoln's Emancipation Proclamation approved by Congress "all slaves . . . are and hence forward shall be free."
April 28	Chancellorsville, Confederate victory
July 4	Gettysburg, Union victory
July 7	Vicksburg, Union victory, Admiral David Porter
Sept. 19	Chickamauga, Georgia, Confederate victory
Nov. 19	Dedication of Gettysburg Cemetery "This government of the people, by the people and for the people shall not perish from the earth."

1864

May 5	Wilderness battles, Petersburg and Spotsylvania, Virginia, Confederate victory
May	Sherman's army takes Chattanooga, Tennessee, then drops south to begin Siege of Atlanta, Union victory
July 2	13th Amendment passed by Congress freeing all slaves
July 2	Congress given authority for post-war Reconstruction
Sept. 4	Atlanta, Georgia, surrender to Sherman's troops, Union victory
Sept. 5	Mobile, Alabama surrender to Farragut's troops, Union Victory
Sept. to Dec.	Sherman's March to the Sea, Union victory
Nov.	Lincoln re-elected. Vice-President Andrew Johnson

Dec. 20	Savannah surrender to General William T. Sherman. Union victory
1865	
March 4	Lincoln inauguration 2nd term "to care for him who shall have borne the battle and for his widow and his orphan."
April 3	Union troops take Richmond Confederacy flees from their capital
April 9	Appomattox, Virginia, Robert E. Lee surrender to Ulysses S. Grant. End of the war.
April 14	Good Friday, President Lincoln shot by John W. Booth
April 15	Death of Lincoln

On August 1, 1863, Lewis Ralston joined the Confederate army. Private Ralston was part of Company 8, 11th Battalion of the Georgia State Militia. Less than two months later the new recruit participated in the Battle of Chickamauga (Cobb County) on September 19-20, 1863. It was the first Civil War battle on Georgia soil. The Southern men forced the Union troops to retreat north. Lewis was 59 years old. [10]

Miles away, on November 1, 1863, three months after his father's enlistment, Lewis Ralston Jr. from Mt Vernon, Missouri, joined the Union army. He spelled his name Lewis *Rolston*. He was 26 years old. Lewis Sr. and his son may never have known they were fighting each other. [11]

Men of the Southern states were enraged by the 1864 action of congress which freed the slaves of the Southern states. Throughout the South some of the newly freed men fled north assisted by Union sympathizers called the Underground Railroad. Some of these newly liberated men joined the Union army.

Before the fall of Vicksburg, which seemed likely in 1863, Lincoln had placed Admiral David Porter in command of the siege. He called General Ulysses S. Grant back to Washington D. C. Grant was named Lieutenant Commander of the Army, later elevated to Commander in Chief. Lincoln then called for loyal citizens to respond to the need for greater effort, his words inspiring 500,000 new army recruits. This action sparked a great increase in volunteers for the Confederate army, strengthening the troops directed by General John B. Hood. Lewis Ralston and all of Georgia knew that when Vicksburg fell, as it did on July 7, 1863, their state would become the prime target for the Union army. Lewis Ralston enlisted 24 days later on August 1.

One of Grant's first actions was to mount a plan which forced Lee's soldiers, who were threatening Washington D.C., to retreat south of the James River. Gettysburg was won by the Union on July 4, 1864, with great loss of life for both North and South. The Gettysburg Cemetery was dedicated on November 19. General William T. Sherman was ordered to leave the Vicksburg area, move to Tennessee, take Chattanooga, and proceed into Georgia.

Sherman's experienced troops reached the Atlanta area in May of 1864 after engaging in skirmishes with the Confederate army under General Hood. The Confederates valiantly defended Atlanta, but finally were forced to abandon the city and flee north, hoping that Sherman's forces would follow. Part of Sherman's army circled the city beginning a massive siege. Other Union men ranged the countryside foraging for food. Dahlonega, 50 miles away, was terrorized by the raiding parties.

Sherman sent Captains George Thomas and John Schofield north in pursuit of General Hood's forces. The remaining Union forces began the overwhelming siege of Atlanta. Some of the Union soldiers were directed to ride through the surrounding area, collecting food and supplies. Atlanta surrendered three months later on September 2.

Sherman's troops vandalized Atlanta. The railroads were destroyed and tracks were pried from their rail beds. The roundhouse, machine shops and all warehouses were burned. Major supply depots were sacked. Any building that might benefit the Southern army was torched. The "set" fires touched off other buildings. Some historians say Atlanta was burned to the ground. Other accounts say that much of the city was spared. One historian wrote: "Fewer than 400 of the city's 3,700 buildings were left standing." (12)

The farms of Lumpkin County had been raided over and over through-out the three-month-long Atlanta siege. Elizabeth Ralston and her children sadly watched as their peaches, ready for harvest, were stripped from the trees. Throughout the counties that ringed Atlanta, cattle were driven off, chickens seized and gardens purloined of fresh vegetables. Wagons and buggies were confiscated to carry the food to Sherman's troops.

On October 9, 1864, Sherman wrote to Grant asking permission for a new course of action, saying he would "make Georgia howl." Grant agreed to the plan and Sherman split his 62,000 experienced men into two wings and prepared to march from Atlanta to the sea. Sherman knew his rested troops could march at least 12 miles a day but this was faster than the supply wagons could travel. He planned to have his men "live off the land," and reports indicate they did indeed. Sherman's army marched out of Atlanta on November 16, leaving a burning city behind them.

Sherman divided his troops sending one wing northeast under General Slocum and the other wing south under General Howard. Selected soldiers were assigned to search the country for food to provide for the swiftly moving troops. The two wings traveled through the heart of Georgia, between them ravaging a path 60 miles wide. Along the way they met small groups of Confederate soldiers that bravely struggled but could not halt the progress of the Union troops. Lewis Ralston may have been a soldier in one of these small groups. Sherman had instructed his generals to burn a few barns along the way so the smoke would advise him of their location, but more than barns went up in flames. As Sherman's troops advanced, slaves fled from the plantations, some falling in at the rear of the Union army, others fleeing North to freedom. On December 24, the march ended with the capture of Savannah and Fort McAllister, both located on the shores of the Atlantic Ocean.

Early in the new year of 1865, Congress passed the 13th Amendment which freed all slaves. In his second inauguration speech on March 4, 1865, Lincoln called for "malice toward none and charity for all." Richmond fell to Union troops on April 1 and the war ended with the surrender of Robert E. Lee to Ulysses S. Grant at Appomattox, Virginia on April 9, 1865.

President Lincoln and his wife made arrangements to attend a play, *Our American Cousin* at Ford's Theater on Good Friday, April 14. He was fatally shot by an actor, John Wilkes Booth. An accomplice of Booth's named Lewis Paine attacked Secretary of State William Seward the same night. Paine was part of a group assembled by Booth which included other enraged former Confederates. Booth was tracked to a barn in Virginia, and 11 days after Lincoln's death, he was shot and killed by government officers on April 26.

Lincoln had hoped to welcome the Southern states back to the Union with a minimum of punishment. Congress violently disagreed, and with

Lincoln's death, congressional leaders were free to put punitive ideas into action. Georgia was one of the states placed under military rule and the citizens of Lumpkin County became prisoners in their own homes. All adult males were forced to take an Oath of Loyalty to the United States of America.

Dahlonega was named a military post soon after the close of the war. The Presbyterian church became the command center. As more soldiers were stationed in Lumpkin County the Mint building, built in 1838 and closed in 1861, became the barracks for these Union soldiers. The basement of the former Mint was used as a dungeon to hold Lumpkin County citizens who were arrested on the slightest provocation. One Sunday evening a prisoner escaped. Many Dahlonega residents were attending church and they were ordered to form a line and leave the church. Soldiers with fixed bayonets herded them to the courthouse. Hours later they were finally allowed to return home. They found that soldiers had searched their dwellings "running their bayonets through the beds and tearing things up in general." Another incident involved the town marshal who had to flee the town and hide in friendly county homes because he defied the orders of some Yankee soldiers. (13)

Federal soldiers stationed in Dahlonega still continued to supplement their army rations by raiding the farms of Lumpkin County. The economy of the county, as well as that of the entire state, was in ruins. Reconstruction policies divided the large plantations into small farms on which only a small field of cotton could be cultivated. A few of the larger cotton farms were spared, but there were no workers and the profits were few. Some of the smaller holdings would barely sustain a family. Atlanta's railroads were vital to Southern recovery and by 1868 the city had been named the state capital and many rail lines and bridges had been rebuilt. New buildings rose from the ashes of the old and slowly Georgia and the South began to recover.

In 1870, Georgia was allowed to re-enter the United States after the state legislature ratified the 15th Amendment, the act granting African-Americans the right to vote. This amendment was state and federal law, but a few Southern men took matters into their own hands and Ku Klux Klan meetings were held in Georgia. The Klan grew in numbers throughout the Southern states. Members of the Klan terrorized former slaves when they tried to exercise their voting privileges. (14)

Private Lewis Ralston mustered out of the Confederate army at war's end in 1865, and returned to his depleted home. The families of Lumpkin County struggled to replant and restock their farms. Others sacrificed to rebuild their stores and banks. The Ralston family had lived in Lumpkin County for 40 years but Lewis felt life simply had to be better somewhere else. By 1870 he had moved his family 60 miles northwest to Whitfield County. (15)

CHAPTER X

❧ AFTER THE WAR ❧

The Ralston family moved from Lumpkin County leaving their friends of 40 years and their cherished home on the Chestatee River. It seemed another defeat for Lewis Ralston. They certainly could have joined their son and Elizabeth's family in Cherokee Nation West (Oklahoma), but perhaps the contact with Lewis Jr. had been lost. Their new home in Whitfield County was 80 miles from Dahlonega and almost 100 miles from Atlanta. In 1870 traveling 30 miles, either on horseback or by mule-powered wagon, was a long day's travel. The Ralstons moved to what to them was unknown territory. Whitfield County was an escape from the petty tyranny of the federal soldiers stationed near their old home. The soldiers were not re-called from Georgia until 1877. We can surmise that Lewis, ostracized by his marriage to a Cherokee woman, was stubbornly determined to exercise his right to own Georgia property.

Lewis did have a brother, Oliver, living in Gilmer County which lies just west of Lumpkin County. Did Lewis even consider moving there? Brian Hood, a descendant of Oliver P. Ralston, has shared his careful research, and some of his information is the basis of the following paragraphs. (1)

Robert Ralston, the grandfather of Lewis and Oliver Ralston, was born in Scotland where he married Frances Tate. There is no record of their immigration to America. Robert's son, John Tate Ralston, who may have been born in Scotland, traveled the long miles to South Carolina where he married Lettie Harris and where their son, Lewis Ralston Sr. was born in 1804. Lewis' youngest brother, Oliver, was born in Lumpkin County in 1831, three years after the Parks' gold discovery in 1828.

Lewis Ralston's paternal family – research Brian Hood

John Tate born	1794	Clarissa	1816
David	1800	Ruthie	1816
LEWIS	1804	Samuel	1823
Robert	1807	Oliver	1831

Lewis was 27 years of age when Oliver was born, had married Elizabeth and started a family of his own. Perhaps Lewis and Elizabeth visited his family when they all lived in Lumpkin County before the 1839 Removal. Hood has diligently searched for burial sites for his ancestors, both in Gilmer and Lumpkin counties. He found no gravesites for John Tate and Lettie Ralston, nor for Lewis and Elizabeth Ralston. Brother Oliver and his wife Nancy (Colbert) Ralston, along with sister Ruthie, are buried in the Mt. Pleasant Baptist Church cemetery in Gilmer County. [3]

Hood states "John Tate Ralston (father of Lewis) owned a gold mine in Lumpkin County." As has been previously noted, the State of Georgia attempted to destroy all Cherokee records when the state annexed the lands of the Cherokee Nation. Most of the few historical records that did survive credit ownership of the gold mine to Lewis Ralston, not his father. Benjamin Parks is honored as the first to discover gold. Lewis Ralston was quick to follow his example.

One of the few existing documents is the old record published by Don L. Shadburn titled *Cherokee Planters in Georgia 1832-1838* and it includes this statement about Lumpkin County: "Among the best known creeks in the county . . . Bob Ralston's branch." The Scottish ancestor Robert is not known to have traveled to America. Lewis' brother Robert was not born until 1854, more than 26 years after the Ralston/Parks partnership followedby Benjamin Parks' gold discovery. Ralston Branch no doubt was named to locate Lewis Ralston's gold mine. [4]

There are no existing records for the location of the Ralston farm in Whitfield County. The 1870 Federal Census states that Lewis Ralston and his family lived near the town of Dalton, District 892, Whitfield County. The value of Ralston's real estate was listed as $500 and his personal estate as $500. [5]

From the census: [5]

Lewis	age	65	Amanda	21
Elizabeth		57	Zachariah	19
Frances		40	James	18
Nancy		34	Martha	17
Agnes P.		26	Robert	14

The census included the sentence: "How many attended school during the year?" None of the children had done so. The energy of the family had been spent in the move and school was not a priority. Another factor may have been that Lewis was in no financial position to help pay for a teacher. In 1870 some rural schools were organized by a few families who lived relatively close to each other and who together agreed to pay a salary and board a teacher. Such schools were usually in session only a few months during the year.

The census asked how many in the family could read and all the names were checked as readers. The census asked how many could write and check marks indicated that Frances, Amanda, Agnes and Martha could not. From the census we assume that the four girls had learned to read at home, but had not attended school to learn to write. The 1870 census also has a section asking for check marks to indicate those who are "deaf, dumb, blind, insane or idiotic." Today we cringe at such labeling and realize that early census enumerators were not careful.

The saddest statistic is the change in economic status. In 1860 Lewis Ralston had $1,800 in real estate and a personal estate of $200. The 1870 census indicated only $500 in both categories. Lewis Ralston Sr. was the only Ralston family member listed as a "citizen of the United States."

We can only wonder how the family of ten proud human beings managed to survive. They may have owned a cow. They no doubt cultivated a garden. Whitfield County was forested and Lewis and his sons may have hunted for game, but no doubt the penalty was severe if hunters were apprehended. It is possible that Elizabeth and the daughters spun thread and wove cloth for garments or for sale. Did Lewis participate in elections? Did they participate in county affairs? There are no records or living descendants to answer such questions.

A great deal of time has been spent in the effort to find a death date for Lewis and a burial site for Lewis and Elizabeth. Cherokee historian Emmett Starr said Lewis died at age 71, "about 1875." Starr stated Elizabeth died "age 86 in 1898." Sandi Smith of the Oklahoma State Historical Society wrote that the Ralston son, Zachariah, reported "both had died in Georgia." Vital Records for the State of Georgia have no listing for either Lewis or Elizabeth. (6)

Historical society records in Georgia and Oklahoma have been contacted for information about Lewis Ralston as well as museums and state agencies. Many cemeteries have been contacted. Descendants of Lewis Ralston have written or visited the author, but no researcher to date has found the final resting place for Lewis and Elizabeth. They lie in unmarked graves, but their names are remembered by the children of their children.

Gideon Blackburn, a Presbyterian minister, compiled a list of eastern Cherokees in 1884. His goal was to enumerate the eastern Cherokees so they and their children would be eligible for membership in Cherokee Nation West (Oklahoma). He listed six of Lewis Ralston's children: Frances, Elizabeth (Emily Elizabeth), Nancy, Agnes, Amanda and James. (7)

No diaries or letters have been found that would detail the lives of the Ralstons. If the children who left the family wrote letters home these pages did not survive. Their son, Lewis Ralston Jr., is the only sibling whose life has been documented.

Lewis Ralston Jr./Louis Rolston (1837-1904)

All of the information about Lewis Ralston Jr. was furnished by Oklahoma sources. None of the sources agree on every detail.

Lewis Jr. married Eliza Postell in Lumpkin County on December 20, 1859. The couple started west with a wagon, two horses and $5.00 in cash. They settled on a farm in Missouri, but almost immediately were caught up in the slavery controversy and the growing national tension.

The Civil War began in 1861. Two years later, on November 1, 1863, in Springfield, Missouri, Lewis Jr. enlisted and became a Private in Company K, 36th Regiment of the Union Army. He spelled his name Lewis *Rolston*. He was part of several encounters on Missouri battlefields and mustered out of service, in Springfield, on July 1, 1865. He left military service only nine days before the surrender of Southern forces at Appomattox, Virginia, on April 9, 1865. (8)

According to Reverend Gideon Blackburn, Lewis Jr. "was a Republican and keeps well informed on the issues of the day." Many different dates are given for the day that Lewis Jr. and Eliza arrived in Indian Territory (Oklahoma). Lewis quickly became aware of the benefits he would receive if he declared his allegiance to the Cherokee Nation and he did so at age 43, on January 8, 1880. He received an allotment of land. (9)

By his military service Lewis Jr. had declared himself as an anti-slavery advocate, and for the rest of his life he cherished his Cherokee heritage. We can wonder if his father, Lewis Sr., back in Georgia, knew that his son was a Republican and had declared himself a Cherokee. The political affiliation of Lewis Sr. is not known.

Lewis Jr. spelled his name *Louis Rolston* on his Cherokee registration and the spelling was used by him and his family from that time forward. The apparent estrangement from his father had been finalized.

The Rolstons took up his Cherokee allotment (it could have been 80 acres or more) in northeastern Indian Territory (Oklahoma). They built a log house and cultivated the land so abundantly nourished by the nearby Grand River. The new dwelling was on the highest point of land and near a clear bubbling spring. The Rolston farm became part of the island when in 1940, more than a century later, the Pensacola dam was erected creating the Grand Lake O' the Cherokees. [10]

Louis Rolston was a prosperous farmer who specialized in cattle production, shipping his stock by rail to St. Louis, Missouri. He was known as "straight forward in all his dealings." [11]

Louis purchased additional acres as his fortunes increased. He gave land for the Needmore School. (Needmore was a small town that adjoined the Rolston property. The author could not find the "more" for which the early town builders had a "need.")

Louis and Eliza also donated land for a cemetery which later was named for them – Rolston Cemetery. Part of this cemetery land was later given to build Lake Center Baptist Church. Their daughter Lou Rolston Lee, who married Needmore's first schoolteacher, was the treasurer and record keeper for the cemetery. Lou and some of her friends dug up small redbud trees, walked to the cemetery cradling the saplings in their arms, and transplantedthese tender shoots on the cemetery grounds. Louis and Eliza had six children: Eliza, Josephine, Frances, Ida, Lou and John. [12]

In addition to his own Cherokee registration, Louis encouraged his brothers and sisters to enroll, and by 1900 most of them were officially documented. In later years John Tate, the oldest brother, and Robert David, the youngest, both lived near the Rolston's Oklahoma farm. Rebecca the oldest daughter married and lived in Georgia.

Nancy Ralston, two years older than Louis (Lewis Jr.) was accepted into the Cherokee Nation on October 18, 1906. At that time she was the widow of William Ransom and was living in Needmore. On her letter of application she said that she previously lived in Georgia and that she took care of her sister Emily Elizabeth "until she died." She wrote that she also cared for sister Fanny (Frances) "who suffered a long time until she died." She listed the names and birthdays of 10 of her siblings, but unfortunately these dates are different from all others. She wrote that her father, Lewis Sr., died on December 1, 1869, but this date is incorrect because Lewis was alive and listed on the 1870 Whitfield County, Georgia, census. She said that her mother, Elizabeth died August 21, 1870. Nancy was 71 years old in 1906 and probably was relying on memory alone.

Louis Rolston (Lewis Jr.) died March 29, 1904, age 67, and was buried in the Rolston Cemetery, as was Eliza in later years. Louis had been attacked by a "mad dog" causing amputation of his leg. Medical treatment of that day could not prevent massive infection. (13)

Last Georgia Years

Lewis Sr. and his family cultivated the Whitfield County farm and hoped for better days. He had taken the Oath of Allegiance to the United States, but all Georgians were still branded traitors by the federal government. Northern Yankees, "Carpetbaggers," rushed to the South to take advantage of Restoration policies. Federal funds had been allocated to find all graves of Union military and remove these bodies to cemeteries in the North. No such funds were allocated to find, identify and mark the graves of Confederate soldiers.

The nation as a whole was steeped in sorrow at the loss of 620,000 young men, both Union and Confederate, who had died in the war. Dr. Drew G. Faust in her book, *The Republic of Suffering,* says that the Union lost 362,222 men, and the Confederacy lost 258,000 – brave soldiers all who would never return from the battlefields. Not included in the total were those 1,094,446 who were wounded. Some of the wounded would never again take an active part in ordinary daily activities. The total of 620,000 deaths was 2% of the total population of the United States (36 states). Dr. Faust writes that 2% of the nation today would represent the loss of six million lives. (14)

The end of the dreadful war in 1865, the realization of the loss of life for both North and South in the struggle, and finally the death of President Abraham Lincoln cast a pall of sorrow across the land. Sentimental stories, poems and songs gave voice to the nation's grief, and the only comfort was the thought of honor, glory and heroism accorded to the fallen military. The great loss of life was justified and clothed in the mantle of patriotism.

Georgia was readmitted to the Union in 1870 and adopted a new constitution in 1877, both granting some equality to former slaves. However, such laws were loosely enforced. Although many large plantations had been sliced into numerous small holdings, the citizens of Georgia still clung to the idea of cotton as the major industry. This idea persisted even though slave labor no longer existed. Gradually textile mills were built increasing the manufacturing base. Georgia farmers increased the production of peaches, pecans, and corn and began to specialize in the development of choice beef.

The development of the peanut industry was far in the future. George Washington Carver, in 1896, at Alabama Tuskegee University, began his research and invention. By 1910 he had created 300 uses for the peanut, and by 1917 the tasty ground nuts were one of Georgia's major industries.

Lewis Sr. and Elizabeth spent their last years harvesting their Whitfield crops in much the same way as when they were first married. They saw little of Georgia's economic improvement.

As has been noted, the best documentation for the deaths of Lewis and Elizabeth came from Emmett Starr who relentlessly researched the history of the Cherokee Nation. His estimate for the death of Lewis is "about 1875." He wrote that Elizabeth died "in 1888." Lewis would have been age 71 and Elizabeth 86 when they died. (15)

Before the Civil War, embalming of the dead was not generally practiced. Large ritualistic funerals were held only for very prominent individuals Many families said farewell to their loved ones with a simple home ceremony. The body of the deceased was washed and dressed in clean garments. The coffin was constructed by family members and within a day, at most two days, neighbors and friends came to the house bringing food and comfort. The body would be laid in a grave on the home grounds in a brief ceremony led by family members. Such a service would have been appropriate for Lewis and Elizabeth Ralston. Their graves, once known by the family, have been forgotten in the passage of years.

Almost everyone marks their days by the inclusion of an historic date such as the beginning and end of wars, the death of a president, or the bombing of our nation's homeland by terrorists. Lewis Ralston had many such events in his life. He had participated in three major nation-changing gold strikes, he had managed to save his family from the upheaval of the Cherokee Removal, he had fought in the Civil War, and he had guided his family to a new beginning in that war's bitter aftermath. Ralston had first-hand knowledge of the nation's history from 1828 to 1875, but there was no one to hear or record his story. Elizabeth was a true partner in all of the family ventures. She birthed and nurtured 16 children. She held the family together when Lewis was prospecting, or fighting his legal battles or participating in the battles of the Civil War.

RALSTON CREEK AFTER 1850 AND OTHER CLAIMS

We can hope that Lewis Ralston knew that his prospecting partners had named Ralston's Creek after his gold discovery on June 22, 1850. He would have been eager to guide Green Russell and others back to his creek in 1858, but perhaps even then he was ignored in favor of Russell. Green had been elected captain of the 1858 return party by the group that included his brothers and cousins. According to the 1858 diary keepers, the party did not even note the date, June 22, when they reached the area of Ralston's 1850 discovery. The exception to other writers was Luke Tierney who stated that they had finally reached "our destination RALLSTON'S CREEK" (sic) in his account of the 1858 return. In later years Green Russell falsely stated that he, Green, had been part of that first gold strike. (See chapter VIII Return to Ralston Creek)

None of the early prospectors, either 1850 or 1858, could possibly have comprehended the impact of their two expeditions. The Russell party founded Auraria in 1858, the first town of any importance in what would become Colorado. The Russells came to the wilderness because of Ralston's gold and they were followed in 1859, only one year later, by hordes of eager prospectors. Auraria's birth was followed by the founding of Denver City later the same month. Residents met in Auraria to plan the beginnings of the State of Colorado.

There have been numerous claims for gold discoveries in what would become Colorado. James Purcell's claim is one of the earliest. He met army lieutenant Zebulon M. Pike in Santa Fe. Pike had been asked to lead an exploring party into the Rocky Mountains to ascertain the scope of the land that had been purchased by the Louisiana Purchase of 1803. Pike and his soldiers in 1806 discovered the grand peak that was named Pikes Peak in his honor. They traveled up into the mountains following the Arkansas River, and built a stockade for shelter near present-day Florence, Colorado.

In the spring of 1807, they strayed south of the Arkansas, unknowingly venturing into Mexico. Mexico had gained independence from Spain and strongly protected its northern border on the river. Pike and his companions were arrested and taken to prison in Santa Fe. In his reports, Pike wrote about Purcell.

James Purcell was a former trapper who enjoyed trading with the Indians in South Park near present-day Fairplay, Colorado. He too strayed into Mexican territory and was taken to Santa Fe. His prison guards found traces of gold in his saddlebags and questioned him at length about his treasure. Purcell did not remember or did not want to divulge the site of his discovery. The Spanish/Mexican soldiers were passionate about the search for gold. They finally decided Purcell's discovery was too vague to justify the effort of a prospecting party. [16]

In 1846-48, the United States and Mexico fought to establish firm borders and the Territory of New Mexico was created by Congress.

Fur trader, George Simpson and others, built small Fort Pueblo in May of 1842. Lavender wrote that Simpson showed some gold dust at Big Timbers in the spring of 1859. Simpson had recruited some of Bent's Mexican workers and they had panned the waters of Cherry Creek. This may be the source of what has been named "Mexican Diggings." It was of course after Ralston's 1850 discovery. [17]

Fort Pueblo, erected in 1842, was a modest adobe fort, and was not well maintained. The inhabitants were traders not gold seekers. Ute Indians destroyed the fort and massacred the few inhabitants in December of 1854. In the 1860s the town of Pueblo began to develop as a trading center for the surrounding farms.

In 1858 a Kansas newspaper wrote about an elderly French trapper named Carriere who said he found gold in the Arkansas River in 1835. No other evidence for his claim exists. [18]

David Lavender wrote about Asa Estes, a "Taos tavern keeper, who often acted as Charles Bent's messenger." Asa was involved in many activities with the Bent brothers, but is not credited with any gold prospecting. Estes Park was founded by Joel Estes, who with his wife and children arrived in Auraria in June of 1859. Later the same year he pioneered the trail to the beautiful high mountain valley later named Estes Park after him. [18]

The Delaware Indian, Fall Leaf, in 1857 showed gold nuggets he had found in the Rocky Mountains, but he refused to lead the Kansas prospectors to the discovery site. These Kansas gold seekers founded Denver City in November of 1858, but Georgia men, following Lewis Ralston's trail, had established Auraria before them.

A recent publication reviews the unpublished manuscript of Marshall B. Cook "written in the 1880s."Cook wrote that he had discovered a gold mining claim marked by an "Estes Party" in 1834. Cook said that he in 1858 could identify the claim by "three to five boulders that marked the corners, being nearly half buryed (sic) in the earth." J. Wendel Cox, Senior Collections Librarian at the Denver Public Library has endeavored to find any record of an "Estes Party." Cook's manuscript is unavailable due to the renovation of the Colorado Historical Society and the only documentation for a man named Estes is noted above. [19]

There is no record other than Cook's for an Estes gold claim. Even today giant rocks and stones regularly tumble from the heights of Clear Creek Canyon. When Cook wrote his account in 1880, he was describing an event that had occurred in 1834, a span of 46 years before. His own gold seeking efforts had happened 22 years previous to his effort to recall the past. [20]

Lewis Ralston would be very proud of the fact that the site of his gold find was placed on the Colorado State Register of Historic Properties on December 1, 1995. He would be amazed to find that Ralston Creek, Ralston Road, parks and business centers carry his name. His small but important gold find opened the door to Colorado's modern history.

Gold Strike Park in Arvada includes fourteen acres of ancient cottonwoods and prairie grass and honors the Ralston discovery. The coming new commuter train that will link Arvada to Denver is named the Gold Line to commemorate Ralston's find.

Modern day descendants of Lewis Ralston have visited the author and some of them did not even know of Ralston's gold discovery in 1850. Some knew but did not realize the historic importance of the event. All were looking for information about their ancestor.

The most complete information about the family of Lewis Ralston comes from Oklahoma:

The Lewis Ralston Family
Ani-Kawi-Deer Clan (Ga-ho-ga)
Courtesy Sandi Smith Oklahoma State Historical Society

	Born	Died
Lewis Ralston, Sr.	1804	1875 about
Elizabeth Kell	1812	1898
Children		
Rebecca (Becky)	1827	1871
John Tate (Jack)	1828*	after 1906
Alexander	1929	Infant death
Frances Tate	1831*	1887
Emily Elizabeth	1833*	1890
Nancy Caroline	1835*	after 1906
Lewis Jr. (Louis)	1837*	1904
Eliza Louisa (Louisa)	1840*	1880
Agnes Paschal (Aggie)	1842*	after 1906
Henry Clay	1844	after 1901
Isaac	1846	Infant death
Zachariah Taylor	1848*	1898
Amanda D.	1849	1854
James David	1850	after 1906
Robert Duncan	1854	after 1901
Martha Josephine	1858	after 1906

* indicates membership in the Cherokee Nation

John Benge – Chief of the Cherokee Deer Clan
Daughter Dorcas married Captain John Lightfoot
After his death she married Charles Duncan
Daughter Emily Duncan married Alexander Kell
Daughter Elizabeth Kell married Lewis Ralston
Son Louis Rolston married Eliza Postell

Gold Strike Park information kiosk

CHAPTER XI

❦ GOLD STRIKE PARK ❧

My acquisition of Lewis Ralston information began in 1970 when I purchased a photocopy of the John Lowery Brown diary. My collection now fills a library storage container. In 1972 I purchased a second copy of the diary to add to the files of the Arvada Historical Society. In the years since I have paid for transportation, copy charges, film costs, and long distance calls. The files I have accumulated represent my research alone. I also donated costs connected with the writing of grants and articles for the benefit of the Arvada Historical Society and the City of Arvada. I was generously rewarded for writing the history of the Shrine of St. Anne. In the years since 1971 the Arvada newspaper has asked me to write articles and I was paid for one of them. There have been no other reimbursements.

Lois Lindstrom

Years and years ago the majestic land that would become Colorado was an unspoiled wilderness of great beauty. The Arapaho tribes called the confluence of Cherry Creek and the South Platte River "our heartland." Ute families left the high peaks and trailed south along the banks of a stream that in 1850 would be named Ralston Creek. Beaver trappers, in the years 1832-1840, built small adobe forts along the Platte River, but these had all been abandoned by the time of Fremont's 1842-1845 expeditions. Bent's adobe castle on the Arkansas River was destroyed by him in 1849. By 1850, in the vast land mass vaguely identified as western Kansas Territory, there were few evidences of men or families from the 30 United States.

1850   June 22, Lewis Ralston panned an exciting amount of gold from a mountain stream that was named that day for him, Ralston Creek. He had $5 equal to 1/3rd of an ounce of gold in his pan. His discovery was documented by a member of the party, John Lowery Brown. In 1850 an ounce of gold was valued at $16. In the year 2011, 1/3rd of an ounce of gold sold for approximately $681.   - 2043

Colorado's first town, San Luis, was colonized (1852) in the area that would become southern Colorado Territory. These families from Mexico were not empire builders. William Bent, after he destroyed his large fort, moved to a much smaller adobe compound named Big Timbers near the confluence of the Arkansas and Purgatoire rivers. The 1850 McNair wagon train made a brief "lay bye" at Fort Pueblo. On December 25, 1854 the few inhabitants of the post were killed by Indians and the structure was destroyed. There were no towns or villages to greet the 1850 Ralston/McNair prospectors.

1858   Ralston served as guide for Green Russell's prospecting party, and the group arrived at the Ralston's gold discovery site on June 25. Little gold was found and most of the disappointed men, including Lewis Ralston, started home only nine days following the July 4 rendezvous. Thirteen men, who persevered when the others turned back, found paying gold in Little Dry Creek (present-day Englewood). They then moved to the confluence of Cherry Creek and the South Platte River where they built a double log cabin. The cabin served as a meeting place and a rustic home.

        November 1, 1858 the Russell men wrote a charter for the Auraria Town Company, so-named in honor of the nation's first gold discovery near Auraria, Georgia in 1828. Reports indicate this charter was carried to the Kansas Territorial Capital at Topeka. The records in Denver are the only ones that survive. Soon a group of prospectors from Kansas Territory arrived carrying organization papers for their town. They erected their tents on the opposite bank of Cherry Creek, facing Auraria. On November 22 they named their settlement Denver City, after the retiring Kansas Territorial Governor, Willliam Denver. The two small towns united under the name Denver City in 1860.

1859   By the spring of 1859 major gold discoveries in the mountains had brought a flood of eager prospectors to Auraria/Denver and to the Ralston Creek/Clear Creek confluence. Historians began trying to document the beginnings of this gold rush in far western Kansas Territory. They could find no living person to answer the question, "Why did gold seekers come to the Rocky Mountains in the first place?

1860   Historians Jerome C. Smiley and Ovando J. Hollister wrote of a man named Sage who said he had found gold in the Rockies.

Hollister wrote about a party of Cherokees finding gold on Ralston Creek in 1852. R. B. Marcy wrote of finding gold in the Platte River. Hurbert Bancroft wrote about an Indian, Fall Leaf and his nuggets "tied up in a handkerchief." None of these early discoveries had a firm date or location. Most early historians simply began their accounts of Rocky Mountain gold with the Russell discoveries. None of the members of the 1850 Ralston/McNair party had remained in the area. They, like Lewis Ralston, had returned to Georgia and Oklahoma. John L. Brown with the 1850 party wrote a diary but it was carried back to Tahlequah and then lost.

Green Russell was quoted as saying that in 1849 he was part of the first gold discovery on Ralston Creek. There are no records for either of the two prospecting parties that Green Russell led to California. One of his descendents wrote a book about Green attempting to trace his wagon tracks. Unfortunately she copied almost word-for-word from the 1850 John Lowery Brown Diary and simply ignored the fact that the Russells were in California in 1850, not on Ralston Creek. Lewis Ralston, as documented in Brown's diary, *was* on Ralston Creek. Dr. Levi Russell, Green's brother, told Colorado Historian LeRoy Hafen that they learned of Ralston's discovery after they, the Russells, returned to Georgia from their second California trip. There are no records for a Russell prospecting group starting west from the Grand Saline (Oklahoma) in either 1849 or 1850.

1861   Colorado Territory was created by the U.S. Congress on February 28, 1861. The civic-minded men who petitioned for the territory met in Dr. Levi Russell's rustic cabin, in Auraria, on the banks of Cherry Creek. Dr. Russell had moved from the first log house and into his own dwelling.

1862   The 1861 United States Land Survey team under the leadership of William Ashley reached the area of Ralston's gold find and recorded: "July 2, 1862. Land from the base of the bluffs north is rich bottom, nearly level and partly cultivated. Timber only found on the banks of the creeks. Cottonwood, boxelder and willow. . . There are gold mines which are worked and pay from $1 to $3 per day to the man." (In 2011 Ashley's $3 in gold would be enormous pay for a day's work. (Records Vol. E page 10)

1870 The Colorado Central Railroad was built from Denver to Golden. The rails reached the high bluff called Ralston Point in September, and Benjamin Franklin Wadsworth founded the town of Arvada. The account of the new Town of Arvada was publicized on December 1, 1870 – the official founding date.

1876 Colorado Statehood on August 1, 1876.

1951 Arvada, population, 2,359 in 1950, was named a "second class" city on October 31, 1951. The population would explode in the next ten years, reaching 20,010 in 1960. During these years Arvada purchased small parcels of land in the gold discovery area: September 1936, May 1951, October 1956, and November, 1958. The land was used for expansion of the sewer treatment plant, the dog pound and other such amenities.

1970 Lois Cunniff Lindstrom read a typed reproduction of the John Lowery Brown diary she found in the files of the Colorado Historical Society. In 1970, more than 120 years after Ralston's gold find, his story was only a footnote in written Colorado history. Historian Muriel H. Wright copied the diary exactly as written by Brown. She researched place names and identified some of the prospectors who were with the Ralston/McNair party. Her valuable work was published by the Oklahoma Historical Society in 1934, but seemingly was not known or thought worthy of interest by Arvada or most Colorado residents.

1971 Lois met with Arvada City Council in October. The Council was holding regular sessions in the old building on Grandview Avenue, and their main concern was an effort to build a badly needed new city hall. Lois asked them to direct some attention to Ralston's gold find and to provide leadership toward preservation of Arvada's history. No action was taken.

1972 After Lois finished the requirements for her master's degree she wrote a pageant for the sixth grade students at Secrest Elementary School. *Look Back With Pride* was presented on two evenings in April of 1972. One of the scenes was a depiction of Lewis Ralston's gold discovery. At the grand finale of the pageant, on both nights, the audience cheered at the mention of an Arvada museum.

1972   August 24. The Arvada Historical Society was organized. Founder Lois was elected as president.

1973   *Waters of Gold* was published by the Arvada Historical Society. Lois wrote the Introduction which included the Lewis Ralston discovery.

1973   The first meeting of the Arvada Cultural Center Committee was held on June 4 in meeting room B at Arvada City Hall. Lois and Councilman Kris Kriofske had been appointed by Arvada City Council to assemble a group directed to study the feasibility of a community center. The two contacted representatives of many civic organizations Lois was elected chairwoman on July 11. A presentation was made to City Council and the group was given permission to hire an architect.

Harold Carver was selected from a panel of thirteen applicants. A history museum was an important component of the final plan developed by the committee and architect Carver. City Council decided to hold a bond election to raise funds to build the proposed center. The ballot had a second section which would provide funds to develop parks and trails. The Arvada Historical Society and many other volunteers, including the supporters of the parks initiative, donated hours and dollars in the effort to educate Arvada citizens about the benefits of the two proposals. The Cultural Center Committee raised funds by presenting a sampler of entertainment from ballet to blue grass that was titled *Forecast*. The election was set for 1974.

Lois and W. A. Lindstrom, "Lindy," met with members of the Arvada city staff. The Lindstroms asked for some recognition or dedication of Ralston's gold discovery site. They received no encouragement. A major problem was the fact that much of the land in question had not been annexed by the city, and Arvada owned only a small portion of the land at the confluence.

1973   Lois wrote an application nominating the Arvada Flour Mill to the National Register of Historic Places. The honor was received in 1975.

1974   May 21. The bond election allowing Arvada staff to issue $3.6 million in bonds to build the cultural center passed successfully. The winning margin was 273 votes. The vote on funds for parks and trails also passed handily.

The once beautiful banks of the Ralston/Clear Creek confluence had become an eyesore. It was the location of the settling pond for the sewer treatment plant. It was the place where the police department stored impounded cars. It had been the site of the dog pound and many cats and dogs had been carelessly buried near the pound structure.

Other debris consisted of unwanted items deposited by city residents. An old, but not historic, mobile home court occupied the area closest to Clear Creek on Jefferson County land not annexed by the City of Arvada.

1975   The Lindstroms wrote an application nominating Ralston's gold site to the National Register. On April 15, 1975, the application was mailed to Washington D. C. Unknown to them, the Arvada City Attorney had written a letter to the National Register stating that Arvada did not want designation because there were sewer pipes in the area. A letter had also been sent, probably at the direction of city staff, from the Historian of the State Highway Department. The historian stated that there was little validation for a Lewis Ralston Gold Site designation.

1975   January 15. Ray Printz, Director, and members of the Jefferson County Open Space committee met with the Lindstroms, Bruce Buell and other members of the Arvada Historical Society at the Lindstrom home. James Hartmann, President of the Colorado Historical Society was also present. Both Printz and Hartmann pledged their individual support but everyone realized that cooperation from the City of Arvada was mandatory.

January 20. Lois, backed by many members of the Arvada

Historical Society, made a presentation to Arvada City Council. She prepared a report detailing the history and tentative plans for the Ralston gold site.

April 28. Arvada City Council passed resolution No. 75-25. The resolution recognized the historic importance of Ralston's discovery and designated the creek bank and a few feet of ground on either side of Ralston Creek as the Lewis Ralston Gold Discovery Site.

May 1. Information was sent by Arvada city staff to the Colorado Historical Society defining the then boundaries of the Lewis Ralston Gold Site. The application to the National Register had been endorsed by the State Review Board, but it was obvious that the National Register would reject the application, and Lois wrote to them withdrawing the nomination application. Lois felt there would be a more favorable time to try again.

The Arvada Historical Society was granted permission by Arvada City Council to hold a recognition ceremony at the gold site on June 22. This was considered to be a "kick-off" for Arvada's participation in the nationwide Bicentennial Celebrations held in honor of the Nation's 200th birthday on July 4, 1776.

1975   June 22. Dedication of Lewis Ralston Gold Site.
The dedication was planned for 2 p.m. Arvada city staff was very helpful. They provided a dump truck on several occasions to hold the refuse of the Arvada Historical Society clean-up efforts. The city also provided a power pole so that electrical tools could be used. The city moved in a large handsome boulder that was planned to hold an historic plaque.

Volunteers under the direction of Lindy Lindstrom built a stage from used lumber and mounted large drawings as a backdrop. Lois stitched large strips of red, white and blue cloth together to make bunting. Ann and Harold Storm assisted in draping the stage. The State Highway Department donated a large sign. Campfire girls posted the colors and a 4-H group planted flowers. Historical society volunteers, directed by Marilyn Lytle, borrowed folding chairs and placed them in rows for spectator seating.

The program included the Arvada High School marching band and the Jefferson Civic Chorus, Campfire girls, a chorus from Arvada West High School, and a 4-H group. Speakers included Donald Feland, Mayor of Arvada, Harry Parmenter, Arvada Chamber of Commerce, Bruce Buell and Lois from the Arvada Historical Society and James Hartmann from the Colorado Historical Society. Hartmann spoke of a marvelous plaque being prepared by the Colorado Historical Society that was not finished. It was presented to the Arvada group at a later date and is currently displayed in the atrium at City Hall. The Arvada Chamber of Commerce hosted refreshments at the close of the ceremony. It was an event greatly enjoyed by the large and interested audience.

1976   July 4. Dedication of the Arvada Center for Arts and Humanities (The Arvada Cultural Center)

1981   Lois and Lindy Lindstrom drove to Tulsa, Oklahoma to tour the Gilcrease Institute of American History and Art. Here they studied the original John Lowery Brown journal, most especially the entries telling of Lewis Ralston's gold discovery.

Bruce Buell, Vesta Miller and Lois met with Craig Kocian and Mary Fran Allen, the latter members of Arvada city staff. Attorney Buell had prepared a legal brief, signed by members of the Arvada Historical Society, which stated that the city would be allowed to stipulate conditions to a National Register application. This would allow the nomination to proceed. His brief stated that for example, absolute access to all sewer lines would be allowed. City staff said that the Buell proposal would be studied. No action was taken.

Bruce Buell and Lois met with city staff members Ron Culburtson, Craig Kocian and City Attorney Ben King. King had prepared a statement:
"August 12, 1981
If the Ralston Creek gold site is listed on the National Register for Historic Sites the City may encounter intolerable delays whenever they seek to undertake a project . . . There seems to be no exceptions in any of the applicable legislation . . .

or other exceptions which would make any foreseeable project of the City more tenable or permissible under existing laws."

Lois, with approval from the Arvada Historical Society decided to try for nomination to the State Register of Historic Properties and to table any try for the National Register.

Lois met with City Manager Neil Berlin, Assistant to the City Manager, Mary Fran Allen and Director of the Planning Department, Ron Culburtson. She asked for cooperation in preparing an application to the State Register of Historic Properties and work on the application began.

1983   June 20. Report to City Council.

Arvada City Council passed Resolution 82-34 which authorized the use of Open Space Funds to purchase property which could be designated as the Ralston Creek Gold Discovery Site.

April 25. Lois prepared a report for Mary Fran Allen, Assistant City Manager, and Ed Talbot, Director of Community Development -Block Grants.

1984   February 4. Presentation to Arvada City Council regarding designation of a park marking Ralston's discovery. Lois prepared and made copies of the report.

July 26. Wallace A. Lindstrom died.

1985   Arvada City Council passed a resolution reaffirming the designation of the Ralston Creek Gold Discovery Site.

1986   Lois was notified that Arvada staff had prepared a C.P.I and that the first steps had been implemented for a gold site park.

1985-86 Lois and Jim Hunt, Drama teacher, Arvada West High School, produced a VCR tape named "The Ghost of Ralston Creek" written by Lois. Local residents and students enacted the Ralston gold discovery performing on the actual banks of Ralston Creek. The tape was filmed by Jim. Lois paid all the initial expenses. The historical society later made copies which were sold to finance society activities.

Arvada City Council allocated $5,000 in the annual budget for a gold site park.

1986   June 22. Designation (second time)
Arvada's staff did an amazing job of cleaning and beautifying the site and providing a power pole for electricity. Lois worked with the Wheat Ridge Recreation District to rent their show wagon to serve for a stage. Rental fee was paid by the City of Arvada. Some of the bows from the first dedication were used to decorate the stage and flowers were donated by local florists. North Jeffco Parks & Recreation provided fold-up chairs and the Arvada Chamber of Commerce set up decorated tables for refreshments. A large crowd thoroughly enjoyed the program which featured an address by Colorado Governor Richard Lamm. He was present only because, he told the crowd, "Lois talked to me." Everyone enjoyed the music of the Ensemble singers from the Arvada Center Chorale who sang the Colorado state song. Featured speakers included Bruce Buell and Edna McCormack from the Arvada Historical Society, Ron Sammons from the Olde Town Arvada Association, Michael Shannon, President of the Arvada Chamber of Commerce, Rex Lowell, from the North Jeffco Board and James Hartmann from the Colorado Historical Society. Mayor Robert Frie dedicated the park and Lois gave a brief history of Colorado's first documented gold discovery.

Everyone enjoyed the refreshments and some of the group later drove to the Arvada Center for the Arts & Humanities to enjoy a reception honoring Governor Lamm.

It was a wonderful day! However, the legal boundaries of the site had not been established, and this most important spot of land was not listed on the State Register.

1990   Colorado voters successfully passed an initiative to allow gambling in Black Hawk, Central City and Cripple Creek. The new constitutional amendment was offensive to some but many taxpayers were persuaded because 28% of all casino profits were to be devoted to historical preservation.

The Colorado Historical Society was chosen, and rightfully so, as the administrator of these funds. Fund personnel developed the forms by which petitoning grants would be evaluated. They were explicit and demanding, and funds are correctly admistered. All projects must be listed on the State Register of Historic Properties. The Ralston site was not because the application to the National Register had been discarded. Lois took up the work.

1991   The State Register (first step before applying to the Fund) was inundated with applications. The Committee developed stricter requirements for all projects, and would accept no new appliction until new forms were printed. It was months before Lois received the packet and could begin a new try for recognition for the gold site.

1992   Lois was asked to serve as Secretary to the Board of Directors for Forward Arvada Building Corporation. She owned no property in old Arvada, but they knew of her dedication: Dr. Galen Callender, President, and R.W. (Bill) Ashton and Dwight Griggs Directors. FABC was determined to focus attention of RTD (before FasTracks) on some form of rail transportation for Arvada. Commuter rail went to Five Points and other lines were planned for south Denver. Lois wrote articles, made telephone calls and organized group excursions to RTD meetings. She made large badges featuring gold pans for attendees to wear. Arvada's delegation was visible - even Mayor Frie attended once. Some nights she and the group were not called on until after 11 p.m. FABC leadership terminated after the RTD election of Terri Binder in 1995. She was followed by Dick Seargent, Wally Pulliam (ten years) and in January, 2011 by Lorraine Anderson. Ground breaking for the Gold Line was held August 31, 2011.

In 1994 Councilwoman Shelley Cook asked Lois to accompany her to a regional RTD meeting. Tentative studies by RTD were calling possible Arvada rail "The Beer Train." Lois convinced those present that Arvada's tracks should be called "The Gold Line."

1992   Arvada City Council planned a project that would extend historic Ralston Road to Sheridan, a badly needed improvement for traffic easement. The budget also included plans to expand and improve Arvada City Hall and the Arvada Center for Arts & Humanities. Gold Site Park development was included.

Actions of a community group called SARA forced a bond issue election by demanding that citizens vote on such expenditures of tax money. A group to support council's original actions was proposed. Lois and Eldon Laidig were asked to lead the group. They asked former mayor Vesta Miller to serve with them and the three served as the Steering Committee. Lois chaired all the meetings. Many volunteers joined the committee. The greatest hurdle to overcome was the wording of the proposal on the ballot. To answer SARA's charges the voter had to put an X in the "No" box. Because of the wording, a voter in favor of the Arvada Center, for example, had to vote "No." The volunteer group had only three months to explain the ballot to Arvada citizens. The group chose the name PRO-ARVADA. Donations helped fund the preparation and mailing of material. A city-wide effort was chosen for a Saturday and volunteers carried a brochure to almost every house in almost every neighborhood. A large rally was held at the Lion's Club. PRO-ARVADA passed successfully on August 6, 1992. Funding was now available to begin the development of all the projects including Lewis Ralston Gold Site Park.

1992   Lois again began the process of applying to the State Register of Historic Properties. Typing of the application on her old Royal was difficult because the form required frequent adjustment for the uneven spacing of the questions. She photographed Ralston Creek and collected letters of support. The City of Arvada engineering department prepared specific maps to accompany the application.

Lois published *First Gold*. Her continuing research had revealed new information and she wanted the story, first told in *Waters of Gold*, to be correct.

September 1. The application to the State Register of Historic Properties was hand delivered to the Denver office by Lois. The 40 pages were too bulky for the Lindstrom stapler so Judy Lydick of the city manager's office had put everything together. City

Manager Neal Berlin was interested in the project and asked his staff to assist in any way.

Lois shortly received a telephone call from the State Register director that the application had been received. She was also told that if Readers of the application had any questions she would be notified in order to prepare a thoughtful answer.

1992   November 20. State Review Board
Lois and city staff members Gordon Reusink and Jim Root attended the meeting. They were confident about the outcome because Lois had received not one question about the nomination. Lois, Gordon and Jim were ready and waiting to speak at 1 p.m. Denver was in the throes of a severe winter storm, and the Arvada group was asked to move their scheduled 2 p.m. presentation to 4:30 p.m. This would allow an out-of-town presenter to use the 2 p.m. slot and start home earlier. Lois later learned that all of the out-of-town presenters had made hotel reservations. The Arvada group waited and finally the presentation was made at 5:30 p.m. The three had to drive snow-packed highway I-70 back to Arvada City Hall. Lois then, in her own car had to drive 10 ice-slicked miles to her home in west Arvada. The drive would have been depressing because of the storm but the three were too sad to converse because their nomination had been rejected. Some of the comments of the Review Board were derogatory. Lois did answer all the remarks. She certainly could have been more forceful if the questions, as promised, had been submitted in advance.

1993   Lois and the city staff began working on a third application to the State Register. Lois was the writer and researcher in addition to which she had to personally visit and secure signatures of permission from the property owners in the vicinity of the planned park. She also contacted city and county leaders for letters of support.

In November, Lois talked with Dale Heckendorn, National and State Register Coordinator for the Colorado Historical Society. On the 30th he drove to Arvada to meet with Lois, Jack Raven, Gordon

Williams and R. W. "Bill" Ashton. He said specific park borders had to be established.

A map of the area showing the Ralston Creek channel as of 1850 was a necessary requirement. Ron Culbertson, Planning Department, gave her the name of Donald Cameron of the Denver Water Department. Mr. Cameron told her that all USGS maps are based on aerial photographs, and that these images can be found in Washington D. C. archives. He provided helpful information and an address. He assisted with the wording of the letter that Lois wrote asking for the earliest known images, and she enclosed her check for $50. The information duly arrived for the year 1937. The black and white images were in sections – like a giant jigsaw puzzle. Culbertson helped identify a few of the sections, and at home she finished the assembly, pasting the sections to a large piece of plywood. To her delight Lois saw that the ancient trees were right where diary keeper John Brown said they were back in 1850. Rick Assmus, city mapmaker, developed a scale map of the area to demonstrate that Ralston Creek was still "6 miles" from the South Platte River, a fact Brown had recorded in his diary.

1993   The City of Arvada purchased Valley Venture Mobile Home Park at a cost of $1,575,000. The property was necessary for the grading of the area. A new bridge was built over the railroad tracks, the channel of Ralston Creek was enlarged and the Ralston Creek Trail was constructed. Much of this was demanded by the Corps of Engineers. The park included space where in 1850, some of the 20 wagons with the Ralston/McNair wagon train probably stopped for the "lay bye." Here they found, said John Brown, "good water, grass and timber."

1994   Mayor Bob Frie and Lois met with the Jefferson County Open Space Committee. The same year the City of Arvada received 3.3 acres of land near the gold site from Open Space funds.

Work began on a new (third) application for designation of Lewis Ralston's discovery site. Judy Lydick in the City Manager's office typed the material assembled by Mike Lee and Lois. The massive State Register application was hand delivered by Lydick to the State Historical Society office on August 28, 1994. The purchase of the trailer park was not finalized, and the city manager, representing Arvada, could not sign the application as the property owner. Thus the application could not be considered by the State Review Board at their meeting on September 1.

The negotiation was soon completed and signed, but the next State Review Board meeting was cancelled.

1995   December 1. Meeting of the State Review Board. Mayor Robert Frie and Councilwoman Shelley Cook agreed to attend, and this evidence of support was a decisive factor. Arvada Historical Society members Bill Ashton, Gordon Williams and Jack Raven transported the large USGS map, and a chart, which they held at the proper times during Lois' presentation. Attorney Bruce Buell drove from Colorado Springs to add moral support. Patrick Kennedy photographed the group. Many city staff members sat in chairs along the wall. Each group petitioning for nomination was allowed 10 minutes. Lois spoke for exactly 9 minutes, leaving time for discussion. There were a few easily answered questions and Director Lane Ittelson called for a vote. It was positive! Gold Site Park was nominated to the State Register of Historic Properties!

The group hurried home to join a crowd of citizens celebrating the 125th anniversary of the founding of Arvada by Benjamin F. Wadsworth on December 1, 1870. All of the Arvada Historical Society members who helped with the state designation had also been involved in planning the celebration party. The announcement was a grand birthday present for the city.

1996   Mike Lee, Park Planner asked Bill Ashton and Lois to be part of a small group to develop plans for a Gold Site Park Task Force.

1996   In the two years following the acquisition of the trailer court property the City of Arvada spent an additional $1,847,796 for costs such as legal fees, demolition, clean up and disposal of unwanted items, including abandoned decrepit trailers left by some former park dwellers. Each occupant was awarded funds to assist their move to a new location. A total of $2.1 million was spent in the years 1994-1996. (Jim Root – City Engineer)

1996   In January, Lois finally received instructions and an application form from the Colorado Historical Society. This was the up-dated form to be to be used in applying for funds from the State Historical Fund (the gambling funds awarded for historical preservation). She wrote to Ron Culbertson, Jim Root and Mike Lee of the city staff as well as Councilwoman Shelley Cook to inform them and she asked for a meeting.

Lois contacted Cindy Nasky, a member of the Historical Fund staff and received helpful suggestions as to a funding grant application. She asked to meet with city staff members: Phil Cortese, Judith Denham, Mike Lee and Jim Root - and Bill Ashton of the Arvada Historical Society.

April 15. A meeting was held. Attending were Lois and Bill Ashton from the Arvada Historical Society, and Ron Culbertson, Patrick Dougherty, Mike Lee and Jim Root of the city staff. The group decided to apply for a grant from the State Historical Fund. Lois would be the writer. A great deal of time was spent discussing plantings with the final agreement that in addition to care of the ancient cottonwoods on the site any new foliage added would duplicate plants that were described in the National Land Survey of 1862. Lois had researched the account written by the surveyor.

1996   July 10. Lane Ittelson, Director of the State Historical Fund, drove to Arvada to attend a meeting arranged by Lois. Attending were Shelley Cook and Ken Fellman members of Arvada City Council, Lois and Bill Ashton from the Arvada Historical Society and city staff members Ron Culbertson, Judith Denham, Mike Elms, Walt Kane, Mike Lee, Jim Root and Jeff Simons. Work continued on the funding application.

Mike Lee, Arvada Park Planner and Lois prepared the text for the grant application and it was typed by Judy Lydick of the City Manager's office. Again Lois secured letters of support and provided photographs. Lois asked Marcia Tate to conduct an archaeological survey of the site. The Arvada Historical Society paid Tate's fee, which she most generously had reduced for the Society. R. W. Ashton (Bill) arranged for two signatures from

Arvada Historical Society officers and a check to be written, all in the space of one day. Lois and Bill met with Tate after hours in the dark foyer of Arvada City Hall. This enabled Tate to begin her work the next day and complete it in time for the grant submission deadline.

1996   October 1. The grant application was delivered to the state office. Arvada's application was denied on March 1, 1997.

December 20. Arvada celebrated the grand opening of the beautiful new revitalized Ralston Road. Traffic now entered Arvada from Sheridan Boulevard via an attractive and modern intersection, and the entrance to Gold Site Park was improved.

Mike Lee arranged for a small group to meet and plan the Gold Site Park Task Force. Lois and Bill Ashton were part of this group.

1997   July 16. A group drove to Denver to meet with Lane Ittelson, Director of the Historical Fund, a meeting arranged by Lois. Attendees included Bill Ashton, Jack Raven, Pat Kennedy, Marilyn Lytle Schroeder and Lois from the Arvada Historical Society, Councilwoman Shelley Cook, and Mike Lee and Maria VanderKolk from city staff. Lois asked Director Ittelson to explain the reasons for the rejection of their State Historical Fund application. Ittelson explained that Arvada's match for the grant would have to be increased. Lois asked for copies of the Readers reports (the individuals who study and evaluate each application.) Rather reluctantly, the copies were made for her. She summarized these and used the information when she wrote the second funding application. Lois found that many of the questions asked by the Readers had already been addressed in Arvada' application, but on the second application Lois was able to highlight the information.

Lois solicited ten letters of support from Arvada and Jefferson County civic leaders and these were included with the application. Again Judy Lydick typed the report using information supplied by Mike Lee and Lois.

1997   Mike Lee asked the following to meet and select a consultant/ designer for the park: Councilwoman Shelley Cook, Bill Ashton,

Judith Denham, Greg Batt, Jim Root, Lois Lindstrom Kennedy and Mike Lee. Tina Bishop and Associates was the firm selected - a wise choice because of Bishop's great interest in the project. Lois helped contact Arvada residents, asking them to join the Task Force, and these dedicated citizens met regularly to make decisions about future plans for the park.

1998   February 1. Arvada's application for money from the State Historical Fund was approved. The city received $25,000 to develop a Master Plan for Gold Site Park.

By this time, the city improvements to the area had begun. Grading was extensive to build the new overpass over the railroad, and the entire 14 acre site was named Gold Site Park by Arvada City Council. Because the federal Corps of Engineers had informed the city that better flood control was necessary for Ralston Creek, the channel of the stream was enlarged. Construction was begun on a elegant bridge that would span Clear Creek and link the well-used Clear Creek bike trail (the trail that connected to Denver trails) to the brand new Ralston Creek Trail.

Lindy!   7-26-84

2000   Patrick Kennedy became ill and died October 28, 2002.

City Council decided that the name of the proposed new park should be officially changed to Gold Strike Park. The park had been identified by a variety of names in the prior years. Lois worked with the city staff and the Colorado Historical Society to facilitate the change.

October 27. Arvada celebrated the dedication of the "signature bridge" spanning Clear Creek. The bridge joined the Ralston Creek trail to the well-known Clear Creek trail. The mast of the beautiful bridge can be seen from highway I-70. Many dignitaries attended the event and Lois again spoke of Lewis Ralston and his gold discovery.

November 7. Arvada voters balloted favorably for a sales tax increase of .0025. The increase would provide funds to be used to purchase bonds, and these funds were targeted for parks, open space and trails. Gold Strike Park was part of the package.

Lois talked with Mike Lee about the possibility of better, more historic information at Gold Strike Park, and they began discussions with North Jeffco Directors Bob West and Larry McGinley. (Arvada staff developed city parks, North Jeffco Parks and Recreation District was responsible for routine upkeep.) Bill Ashton of the Arvada Historical Society joined the team. Mike Lee commissioned the firm of Quinby-Clune to work with the committee. The firm presented the idea for a kiosk.

The kiosk, which resembles an ancient miner's, shack, displays side panels which tell the story of the first documented gold find in Colorado, and the connection of this discovery to the Cherokee trail and to the beginnings of Colorado.

2000   In February the City of Arvada published their Gold Strike Park Master Plan Report. On page 25-26 of this report *Waters of Gold, First Gold* and the *John Lowery Brown Journal* are cited extensively – no authors listed. Lois wrote the first two and discovered the third. The report quotes Luke Tierney, Ovando Hollister and Marcia Tate – all researched by Lois. Author's names are listed on the last page of the report.

On page 6 of the report Lois is credited with research and nothing else with regard to the development of Gold Strike Park The last line states that the site was admitted to the State Register in 1997. Wrong. The site was named to the State Register of Historic Properties on December 1, 1995. The unnamed group in the photograph on the same page does not represent the many, many people who helped create the Master Plan. Proper credit is not given to Park Planner Mike Lee for his great effort to save this important and historic property.

The report begins with the Executive Summary:
> "The City of Arvada and community members have long envisioned the creation of a commemorative park at the site of the first documented gold find in Colorado."

A truer statement would be: The City of Arvada has known about the Lewis Ralston gold discovery since 1971.

2004   June 22. The dedication of the kiosk was held in spite of the fact that Arvada had just experienced eight days of rain. Yellow straw was layered over the mud puddles making a colorful ground cover for the kiosk which had been designed to appear aged and rusty.

The ceremony was planned by the city manager's office. Lois invited Georgia Contigulia, President of the Colorado Historical Society and she spoke of warm relationships with the State office and congratulated Arvada for their historic efforts.

Councilman John Malito was the Master of Ceremonies. Charles Hornbuckle, from Olympia, Washington, related to Lewis Ralston, spoke briefly about his family. He had learned of Lois' work and had contacted her several years earlier. He was researching the history of his great, great, grandfather, Lewis Ralston.

Through the years of effort, she focused on Gold Strike Park. Lois had worked on other Arvada activities such as the historic signs in Olde Town, the nomination of Olde Town to both the State and National Historic Registers, and the Arts in Olde Town Program. She attended many meetings concerning the Arvada Center for the Arts & Humanities. Lois wrote articles for the City of Arvada and for the local newspaper. She encouraged Arvada residents to attend sessions of the RTD board and to participate in discussions concerning development of a commuter train to Arvada. Some early supporters referred to the proposed commuter rail as the "beer train." In 1994, Lois suggested instead that the name be the Gold Line in honor of Lewis Ralston's discovery.

Great thanks are due the members of Arvada City Council who through the years directed city staff to acquire the necessary property and provide the funds for the amazing transformation of blighted land into historic parkland. Gold Strike Park is a tribute to the many Arvada residents who by their interest through the years made the park a reality. Gold Strike Park has been a work in progress for 33 years, 1971-2004.

2011   There have been no further developments at Gold Strike Park since 2004. Colorado residents will take notice of the first documented gold discovery in the state only if Arvada residents insist on the recognition.

Challenges for Arvada:

A nomination to the National Register of Historic Places

An exciting gold exhibit in the Museum at the Arvada Center for Arts & Humanities

Curriculum prepared for students of Jefferson County R-1 public schools, suitable for use by all Colorado students

Activities planned at the site to entertain and educate Arvada residents and visitors.

Lois Cunniff Lindstrom Kennedy
November 2011

ACKNOWLEDGEMENTS

Many individuals in the years since 1971 have assisted my research. Some are listed below. I am very grateful to everyone who has assisted my search during the past 38 years.

Marilyn Lytle Schroeder has shared her great genealogical knowledge and supported me in every way. She found sources that I did not know existed.

Mary Kay Connor Speiler, former Editor of the *Arvada Sentinel* newspaper, read and corrected the draft. The book is greatly improved because of her critical review.

Charles Hornbuckle, one of Lewis Ralston's great, great grandsons, and his wife Suzanne were my first Ralston family contact. Their interest and factual support, including presentation of the book *Cherokee Trail Diaries,* has been invaluable.

John Lowery Brown's 1850 diary is the foundation of this work. Muriel Wright's meticulous transcript of the diary made it available for me. I am most grateful to the librarian at the Colorado Historical Society who handed the typescript to me.

The **Gilcrease Museum librarians** of Tulsa, Oklahoma were most generous with their time and assistance, and allowed me to hold and read Brown's original journal.

Brian Hood of Dahlonega, Georgia, descended from Lewis Ralston's brother, Oliver, shared the results of his genealogical research. I was able to contact Brian through Reverend Clifford Allen, present pastor of Antioch Baptist Church in Auraria, GA. Rev. Allen gave me the name and address of Martha Taylor, and she referred me to her nephew Brian.

Sandi Smith of the Oklahoma Historical Society sent detailed and accurate information on Lewis Ralston and his family. Other sources were used, but this information proved to be one of the most valuable sources.

Merita Rozier, Microfilm Editor for the Georgia Secretary of State in Atlanta provided copies of valuable general information on Lewis Ralston.

The **Dahlonega Gold Museum** in Georgia has been one of my primary resources. The late Madelaine Anthony was my reference for many years, and she was tremendous assistance. In the last few years many other librarians at the museum have continued her work, and helped me describe Auraria and Dahlonega through the Lewis Ralston years.

Patricia and Dr. Jack Fletcher and Lee Whitely authored the *Cherokee Trail Diaries*. This is a most complete and accurate compilation of information about wagon trains originating from Oklahoma in 1849 and 1850.

Rose Stauber, Delaware County Genealogical Society in Grove, Oklahoma provided needed information on early Oklahoma settlers.

Jackie Coatney, Delaware County Library in Jay, Oklahoma supplied information on Louis Rolston (Lewis Ralston Jr.).

Jane Johnston Librarian, Standley Lake Library, Arvada, Colorado many times found the answer to my questions and photocopied material and mailed it to me.

Pamela Wineman, Denver Museum of Nature & Science – Denver, Colorado, furnished information on gold in the state.

Heather Whitehead of Arthur Lakes Library at the Colorado School of Mines in Golden, Colorado shared excellent material on the early days of gold in Colorado.

Mike Lee, Park Planner for the City of Arvada.

LIBRARIES:

Colorado Historical Society Stephen Hart Library - Denver, CO
Chestatee Regional Library - Dahlonega, GA
Denver Public Library, Western History Collection - Denver, CO
Dalton State College Library - Dalton, GA
Georgia Historical Society - Savannah, GA
Golden Public Library - Golden, CO
Lumpkin County Library - Dahlonega, GA
Oklahoma State Historical Society - Oklahoma City, OK
Sutter's Fort - Sacramento, CA
Whitfield-Murray Public Library - Dalton, GA
Whitfield County Courthouse - Dalton, GA
Reference desks in California libraries
Vital Records Department - Atlanta, GA had no record of Ralstons.

William A. Montgomery of the Atlanta-Fulton Public Library sent general information and a helpful letter but had no information on Lewis Ralston.

CEMETERIES OF DALTON, GA:

Dalton Cemetery
Murray Memorial Gardens Cemetery
United Memorial Cemetery
Brian Hood has researched other cemeteries. No record of Lewis Ralston or family.

GENEALOGIES:

In addition to Charles Hornbuckle and Brian Hood, the author is grateful for the information supplied by Ralston family descendants:

Mary Davis - Arvada, CO (present address unknown)
Geri Lilly - Midland, TX
Kelly Hogue - Madison, MS

GOLD STRIKE PARK

I owe great gratitude to the **Arvada Historical Society** who supported efforts to gain recognition for Lewis Ralston's first documented Colorado gold find.

Individuals who made a great commitment are listed at the appropriate place in the document.

Thank you to **Shelley Cook** for her support and encouragement.

Thank you **Bruce Buell,** who served as pro bono counsel for the Arvada Historical Society and who deserves very special mention. Even after his office and family moved to Colorado Springs, he continued to be available for consultation. His support and advice was an important factor in the effort to gain state recognition of Ralston's gold discovery.

I am very grateful for the support and expertise of my editors, **David Robison** and **Lisa Langley.**

≈ SOURCES ≈

Chapter I GOLD pp. 9-11

1. Edwards, Ron and Gladstone, James, GOLD
    Crabtree Publishing Co. 2004

2. Bernstein, Peter L., THE POWER OF GOLD, John Wiley & Sons Inc.,
    New York, 2004 p. 247

3. FEDERAL PLACER MINING LAWS AND REGULATIONS,
    Department of the Interior, February 1938. Courtesy The Library,
    Colorado School of Mines, 1500 Illinois St., Golden, Colorado 80401

4. Sachdev, Ameet, THE DENVER POST, "Gold has new luster for jittery
    investors." Section C, January 21, 2002. Price on 11-1-11 U.S. Federal Mint
    Denver, Colorado

5. Chakraharty, Gargi, ROCKY MOUNTAIN NEWS, "Going for the
    Gold, May 29, 2004

Chapter II CHEROKEE LEGACY pp. 13-19

1. Williams, Samuel Cole, EARLY TRAVELS IN THE TENNESSEE COUNTRY,
    Nashville, Tennessee 1928

2. Woodward, Grace Steele, THE CHEROKEES, University of Oklahoma Press,
    Norman, Oklahoma, sixth printing 1979, page 85
    Also Williams, Samuel Cole
    Also Flanagan, Mike, THE DENVER POST, "Written language Sequoyah's
    gift to his nation." March, 1988

3. Woodward, op. cit. pp. 114-116

4. Woodward, ibid p. 98

5. Woodward, ibid p. 114-116

6. Foreman, Grant, INDIAN REMOVAL, University of Oklahoma Press,
    Norman, Oklahoma, 1934, p.31

7. Holiday, J. S., THE WORLD RUSHED IN, Simon and Schuster, New York,
    1981 p. 33

Chapter III LEWIS RALSTON AND GEORGIA GOLD pp. 21-30

1. Family history information from Charles Hornbuckle, Olympia, Washington
    and Brian Hood, Ellijay, Gilmer County, Georgia

2. Cain, Andrew W. HISTORY OF LUMPKIN COUNTY – FIRST HUNDRED
    YEARS, Stein Printing Company, Atlanta, Georgia, 1952

3. Sherwood, Ariel, GAZETTEER OF THE STATE OF GEORGIA, originally published 1837, reissued for Clearfield Printers by Genealogical Publishing Company Inc. Baltimore, Maryland 2003

4. Mitchell, Larry E. "A Really Golden Heritage," Taken from North Georgia Journal, Vol. 2, Number 1, Spring 1985

5. Mitchell, op. cit. p. 7

6. Mitchell, op. cit. and Woodward op. cit.

7. Coulter, E. Merton, AURARIA – THE STORY OF A GOLD MINING TOWN, University of Georgia Press, Athens, Georgia, 1956

8. Map of Lumpkin County. Basic map courtesy Jane Johnston, Standley Lake Library, Arvada, Colorado
   Details Lois Lindstrom

9. Amerson, Anne D. I REMEMBER DAHLONEGA, Chestatee Publications, Dahlonega, Georgia, 30533
   1994

10. Georgia map from Internet Mapwatch
    "Georgia County Map" from the Internet, Courtesy Jane Johnston, Standley Lake Library, Arvada, Colorado
    Details Lois Lindstrom

10. Shadburn, Don L. CHEROKEE PLANTERS IN GEORGIA
    1832-1838, Pioneer Cherokee Heritage Series, Vol 2 p.228

11. Woodward, ibid, p. 158

12. Davis, Mary, Family History, Interview with author 1975
    No current address available.

13. McCallie, S. W. "The Ralston Mine" Material from the Dahlonega Gold Museum, 1 Public Square
    Dahlonega, Georgia 30533

14. Cain, ib. id. p. 125

15. Coulter, ib. id. p. 14

16. Amerson, Anne D., "Jennie Wimmer Tested Gold in Her Soap Kettle." Material from the Dahlonega Gold Museuem, 1 Public Square, Dahlonega, Georgia 30533

Chapter IV LAND OF GOLD AND GRIEF pp. 31-38

1. Smith, Sandi, Oklahoma Historical Society, Research
        Division, 501 Whitaker Street, Savannah, Georgia, 32401
    Davis, Mary, Family History, Former address 6641 West
        52nd Ave., Arvada, Colorado. Present address unknown
    Hornbuckle, Charles, Family History, 7245 118th Ave. S.W.
        Olympia, Washington 98521
    Hood, Brian, Family History, 762 Old Ellijay Road East,
        Dahlonega, Georgia 30533 Martha Taylor of Dahlonega
        gave the author his name
    Stauber, Rose, Delaware County Genealogical Society
        1140 NEO Loop, Grove, Oklahoma 74344
    Rev. Clifford Allen, Pastor, Antioch Baptist Church
        4632 Auraria Road, Dawsonville, Georgia
    Jane Johnston, Librarian, Standley Lake Public Library
        Arvada, Colorado
    Rozier, Merita, Department of Research, State of Georgia
        Department of History, 214 State Capitol, Atlanta, Georgia
    Kelly Hogue, Family History, 100 Fawn Lane, Madison Mississippi
    Geri Lilly, Family History, 4601 Tottenham Circle,
        Midland, Texas. She gives credit to Rosetta Petty, Rogers, Texas

    Those who unsuccessfully searched for death certificates and other
    information – valuable knowledge for any future research:
        Georgia Vital Records, Atlanta, Georgia
        William Montgomery, Reference Librarian
        Atlanta-Fulton Public Library
        Karen C. Handel, Secretary of State,
            5800 Jonesboro Road, Morrow, Georgia
        Evelyn Tasley, Vital Records Whitfield County
            808 Professional Blvd., Dalton, Georgia

2. Smith, Sandi, Oklahoma Historical Society, op. cit.

3. Sherwood, Ariel, GAZETEER, op. cit.

4. Henderson, Daniel, THE HENDERSON REPORT,
    Henderson compiled the report from the 1835 census

5. Montgomery letter in Chapter IV file

6. Book of Valuation, Mary Davis, op. cit.

Chapter IV *continued*

7. Woodward, Grace, THE CHEROKEE op. cit. p. 218 quoting Cherokee
    authority Grant Foreman
    Also
    Elish, Dan, THE TRAIL OF TEARS, The Story of the Cherokee Removal,
    Benchmark Books, Marshall Cavendish, New York

8. McCallie, S. W., Assistant Geologist, Fieldwork Account of The Ralston Mine
    in Lumpkin County
    Information from Dahlonega Gold Museum, 1 Public Square
    Dahlonega, Georgia 38533

9. Morris, Constance, Supervisor, The Library, Dalton
    State College, 650 College Drive, Dalton, Georgia 30720

10. Vinson, Carl, Institute of Government, Historical Maps, Atlanta, Georgia

11. Horan, F. T. THE F. T. HORAN ARTICLE, Courtesy Grady Campbell, M. D.
    P. O. Box 809, Dublin, Georgia 31040 – and Sherwood, Adiel, op. cit. p. 334

Chapter V LEWIS RALSTON AND COLORADO GOLD, pp. 39-50

1. Fletcher, Patricia K. A. and Dr. Jack Earl Fletcher, and Lee Whiteley,
    CHEROKEE TRAIL DIARIES, Fletcher Family Foundation,
    730 Three Crabs Road, Sequim, Washington, 98382
    This book is the very best source of material concerning the gold exploration
    wagon trains leaving from Oklahoma.

2. Hafen, LeRoy R. and Ann Hafen, COLORADO – A HISTORY OF PROGRESS,
    Old West Publishing Co.
    1228 E. Colfax Avenue, Denver, Colorado, 80218 p. 84

3. Fletcher, et al, CHEROKEE TRAIL DIARIES, ibid. pp. 203, 235, 288, and 325

4. Fletcher, op. cit. p. 19

5. Fletcher, op. cit. p. 23

6. Fletcher, op. cit. p. 208   also
    Brown, John Lowery, THE JOURNAL OF JOHN LOWERY
    BROWN OF THE CHEROKEE NATION EN ROUTE TO
    CALIFORNIA IN 1850, Transcribed by Muriel H. Wright
    Reprinted in Chronicles of Oklahoma, Oklahoma City,
    Oklahoma Vol. XII, Number 2, June l939 p. 191

7. Fletcher, ibid, pp. 201

8. Brown, John Lowery, ibid (May 5, 1850) p. 182

9. Brown, John Lowery, op. cit. footnote p. 182

10. Brown, John Lowery, op. cit. footnote p. 186

11. Wright, Muriel H. Forward to her transcript of the Brown Journal, p. 177

12. Lavender, David, BENT'S FORT, University of Nebraska Press, Lincoln, Nebraska, 1954 pp. 334-335, pp.338-339

13. Broadhead, Edward, FORT PUEBLO, Pueblo County Historical Society, Pueblo, Colorado p. 3

14. Information from The Library, Colorado School of Mines op. cit.

Chapter VI CALIFORNIA GOLD pp. 51-59

The map on page 37
Miller, Donald C. GHOST TOWNS OF CALIFORNIA,
Pruett Publishing Company, Boulder, Colorado 1978

1. Wright, Muriel, typescript of Brown diary ibid. pp. 177-213

2. Wright, op. cit. quoting from Willard O. Waters who served as bibliographer for Americana and the Huntington Library, p. 211

3. Gernes, Phyllis L., HIDDEN IN THE CHAPARRAL
Dome Printing and Lithograph 1979, p. 34

4. Gernes, ibid, p. 115

5. Gernes, ibid. p. 8

6. Gernes, ibid. p. 8

7. Holiday, J. S., THE WORLD RUSHED IN, The California Gold Rush Experience, Simon and Schuster, New York, 1981, p. 354

8. Holiday, ibid. p. 303

9. Holiday, ibid. p. 306

10. Gernes, op. cit. p. 42

11. Gernes, op. cit. p. 129

12. Wright, Muriel, Footnotes in her typescript of the John Lowery Brown Diary, op. cit.:

John L. Brown p. 178

McNair, Clement p. 182

Thomas F. Taylor p. 181

Meigs, Return p. 200

Adair, George W. p. 212

Mayes, Samuel H. p. 212

Chapter Chapter VII TENSION AND DEPRESSION pp. 61-66

1. Federal census record for Dahlonega District, Lumpkin County, Georgia, September 24, 1850, W. P. Reid, enumerator

2. Two other lists give information on the Ralston family: Chapman Report of l851 and The Hester roll of 1884
   Neither list is completely correct. The information furnished by Sandi Smith of the Oklahoma Historical seems to be the most accurate.

3. Coulter, E. Merton, AURARIA, University of Georgia Press, Athens, Georgia, Reprint dated 1956 p. 133

4. Coulter, op. cit. p. 113

5. Sherwood, Ariel, GAZETTEER, op. cit. p. 50-51

6. Sherwood, op. cit. p. 198-201

7. Sherwood, op. cit.

Chapter VIII RETURN TO RALSTON CREEK pp. 67-80

1. Brown, John L., ibid
   Bancroft, Hubert Howe, THE WORKS – HISTORY OF NEVADA, COLORADO AND WYOMING,
   San Francisco, California 1890
   Baskin, O. L. HISTORY OF THE CITY OF DENVER, ARAPAHOE COUNTY, COLORADO, Baskin and Company, 186 Dearborn Street, Chicago, 1880
   Dickson, Colonel T. C. EARLY EXPERIENCES OF Col. T. C. Dickson, published THE TRAIL, Society Of Sons of Colorado, March 1911, Vol. III Number 10
   Hafen Dr. LeRoy R. COLORADO AND ITS PEOPLE
   Vol. I, Lewis Historical Publishing Company, Inc.
   New York, 1948
   Hafen, LeRoy R. CHEROKEE GOLD SEEKERS IN COLORADO 1849-1850, published in the Colorado Magazine, State Historical Society of Colorado , May 1938 Vol. XV, No. 3 pages 109 and 191
   Hall, Frank, HISTORY OF THE STATE OF COLORADO
   Chicago, Blakely Printing Company 1890
   Lavender, David, THE ROCKIES, Harper and Rowe
   New York 1968
   McKimens, William, "Letters from Auraria 1858-1859"
   Published in the Leavenworth Times newspaper,
   Dec. 18, 1858 and Feb. 5, 1859
   Montgomery, Mabel Guise and Silvia Peterson,
   A STORY OF GOLD HILL COLORADO, The Book Lode,
   Longmont, Colorado 80501, 1987

Pierce, James H. WITH THE GREEN RUSSELL PARTY
Written in 1885 published in THE TRAIL, The Sons of Colorado,
W. C. Bishop, Editor, May 1921, Vol. XIII No. 12
Tierney, Luke, HISTORY OF THE GOLD DISCOVERIES ON THE SOUTH
PLATTE RIVER, D. C. Oakes and S. W. Smith, Pacific City, Iowa, 1859

2. Smiley, Jerome C. HISTORY OF DENVER, The Denver Times, Times-Sun
    Publishing Company, Denver 1901 p. 184

3. Fletcher, CHEROKEE DIARIES, ibid.

4. Hafen, Leroy R., PIKES PEAK GOLD RUSH GUIDEBOOKS OF 1859,
    Arthur H. Clark Company
    Glendale, California, 1941, pp. 105-106
    (Quoting Luke Tierney)

5. Hafen, LeRoy R., op. cit. note 160 on p. 111

6. West, Elliott, THE CONTESTED PLAINS, University Press of Kansas,
    Lawrence, Kansas 1998 Chapter 5

7. Hafen, LeRoy R., op. cit. pp. 152-153
    Stone, Wilbur Fisk, HISTORY OF COLORADO, Vol. 1
    S. J. Clarke Publishing Company, 1918, pp. 136-140

8. Smiley, ibid. pp. 208-209

9. Smiley, ibid. p. 191 And Stone, ibid. pp. 140-143

10. Whithead, Heather, Librarian, Arthur Lakes Library Colorado School of
    Mines, 1400 Illinois Avenue, Golden, Colorado. Telephone interview 2008
    and
    Henderson, Charles W., MINING IN COLORADO,
    Department of the Interior, Professional Paper #138
    Washington D. C. , Government Printing Office. 1926 pp. 108-109

11. Hafen, LeRoy R., op. cit. p. 142

12. Petty, Rosetta, P. O. Box 76569 Rogers, Texas
    Mrs. Petty compiled Ralston Family History. She said Lewis Ralston's
    brother Samuel was born in 1853 and lived in Gilmer, County. He had
    not been born in 1850, the year of the McNair/Taylor prospecting party,
    and would have been 5 years old in 1858 when the Russell party returned
    to Ralston Creek.
    (The above information was shared by Geri Lillly
    4601 Tottenham Circle, Midland, Texas)

    THE MAP on page 73 was developed by Ben K. Parker, Jr.
    From his 1974 book THE GOLD PLACERS OF COLORADO.
    Courtesy of the Arthur Lakes Library

Chapter IX PRIVATE LEWIS RALSTON pp. 81-92

1. Shadburn, Don L. CHEROKEE PLANTERS IN GEORGIA, 1832-1838
    Cherokee Heritage Series
    Vol. 2, 1989

2. Cain, Andrew W., HISTORY OF LUMPKIN COUNTY, FIRST HUNDRED
    YEARS 1832-1932
    Stein Printing Company, Atlanta, Beorgia 1932 p. 137

3. Blackburn, Gideon, HISTORY OF INDIAN TERRITORY
    Courtesy Rose Stuber, Delaware County Genealogy Society, 1141 NEO Loop,
    Grove, Oklahoma

    also

    TENNESSEE ENCYCLOPEDIA OF HISTORY AND CULTURE, Information
    copyrighted by Tennessee Historical Society, Nashville, Tennessee. Courtesy
    of Golden Public Library, Golden, Colorado

    also

    Delaware County Historical Society, A HISTORY OF MONKEY ISLAND,
    Authored by Thelma Lee and Jim Swinney. Copyright 1987 by the Delaware
    Historical Society, Jay, Oklahoma. Information on Louis Rolston (Lewis
    Ralston Jr.) p. 14

4. Smith, Sandi, Librarian, Oklahoma State Historical Society, Research Division,
    501 Whitaker Street, Savannah, Georgia 32401
    Information from Dr. Emmet Starr, Rare Books reprint Oklahoma City

5. Donald, David Herbert, LINCOLN, Simon and Schuster, Rockefeller Center,
    New York, New York 1995, p. 141
    This excellent resource was the primary source on President Lincoln and for
    the quotations used. It was very valuable for information on the Civil War.
    The author is Professor Emeritus of Harvard University, and has received two
    Pulitzer prizes for his historical writings.

6. Clinton, Catherine, FANNY KEMBLE'S CIVIL WARS, Simon and Schuster,
    Inc. New York 2001

7. Sherwood, Ariel, GAZETEEER OF THE STATE OF GEORGIA, Originally
    published in the State of New York, In 1837.
    Reprinted by the Genealogical Publishing Company, Inc.,
    Baltimore, Maryland 2002

8. Cain, Andrew W., HISTORY OF LUMPKIN COUNTY – FIRST HUNDRED
    YEARS, Stein Printing Company, Stein Printing Company, Atlanta,
    Georgia, 1952, pp. 142-143

9. Cain, Andrew W., op. cit. p. 145

10. National Park Service, U. S. CIVIL WAR SOLDIERS
1861-1865

11. National Park Service, op. cit.

12. McPherson, James M. ATLAS OF THE CIVIL WAR
Oxford Printing Company. Courtesy Jane Johnston
Standley Lake Library, Arvada, Colorado p. 278

also

World Book Encyclopedia, World Book, Inc. 2005
233 North Michigan Avenue, Chicago, Illinois 60601 Vol. 1 pp. 859-860

also

Heidler, David S. and Jeanne T. ENCYCLOPEDIA OF THE AMERICAN
CIVIL WAR, A Political, Social and Military History Encyclopedia
of the American Civil

12. War, ABC-CLIO, 130 Cremona Drive, P.O. 1911
Santa Barbara, California 93116-1911 pp. 1769-1773
also

World Book Encylopedia, World Book Inc. GEORGIA,
233 North Michigan Avenue, Chicago, Illinois 60601
Vol. 8

also

Denton, Paul, "Sherman's March Through Georgia"
From the internet Website: htlp:/ehistory 050 edv
Courtesy Jane Johnston, Standley Lake Library, Arvada, Colorado

13. Cain, Andrew W. ibid pp. 178-179

14. World Book Encyclopedia see Sources 12.

15. Census record for 1870 "Inhabitants in the 19th land Dist. 892th "
Whitfield County, State of Georgia, Post office
Dalton, Georgia June 28, 1870
William Henderson enumerator

Lewis Ralston	65
Elizabeth	57
Frances	40
Nancy	34
Agnes	26
Amanda	21
Zachariah	19
James	18
Martha J.	15
Robert	14

Chapter X AFTER THE WAR pp. 93-104

1. Hood, Brian, Genealogist, Olive P. Ralston family history
        762 Old Ellijay Road East, Dahlonega, Georgia 30533
        "John Tate Ralston and his wife disappeared. There is no record of burial for
        them in Lumpkin County, Georgia."
        "If he (Lewis Sr.) is buried in Whitfield County it must be in an
        unmarked grave."

    and

    White, Marcarcelle, Ex. Sec.. Whitfield-Murray Historical Society, Inc. 715
        Chattanooga Ave., Dalton, Georgia 30721
        "I have searched the cemetery records for Lumpkin, Whitfield, Murray,
        Gordon and part of Gilmer Counties." She found no mention of Lewis Sr.
        or Elizabeth Ralston

    and

    Stauber, Rose, Delaware County Genealogical Society,
        1140 NEO Loop, Grove, Oklahoma. She looked for a will for Louis
        Rolston (Lewis Ralston Jr.) but Rose reported "A folder with
        Louis/Lewis) was found but it had nothing in it."

    and

    Capponi, Debra, Chestateee Regional Library (Lumpkin and Dawson
        Counties) "No record" 342 Courthouse Hill, Dahlonega, Georgia, 30533

2. Hood, Brian, ibid

3. Shadburn, Don L., CHEROKEE PLANTERS IN GEORGIA, 1832- Pioneers –
    Cherokee Heritage Series, Vol. 2, 1989, p. 217

4. Hood, Brian, ibid

5. Federal Census for Dalton, Whitfield County, Georgia
    June 29, 1870, Distric 892

6. Smith Sandi, Oklahoma State Historical Society, Research Division,
    501 Whitaker Street, Savannah, Georgia 32401

7. Hester Roll, courtesy Rose Stuber, ibid

    also

    Family Tree Maker's Genealogy: User Home Page book:
    Cherokee Lineages: Register Repo

8. U. S. Civil War Soldiers 1861-1865 National Park Service,
    The Generations Network, Inc.

9. Blackburn, Gideon, HISTORY OF INDIAN TERRITORY,
    Louis Rolston Jr. pp. 873-874 Courtesy Rose Stauber

also

Bowen, Jeff, CHEROKEE CITIZENSHIP COMMISSION DOCKET, 1880, Oklahoma State Historical Society

10. Delaware County Historical Society, HERITAGE OF THE HILLS, 10TH Anniversary, 1985, Articles featuring Ruby Jewell Muskrat Mason and Thelma Muskrat Lee both descendants of Louis Rolston.

11. Lee, Thelma and Swinney, Jim, A HISTORY OF MONKEY ISLAND, information on Needmore, Rolston family and the Rolston Cemetery, pp. 11 Courtesy Delaware County Historical Society

12. Lee, Thelma and Swinney, Jim, ibid

13. Smith, Sandi, op. cit. Information on Nancy Ralston Ransom

14. Anderson, Jimmy E. "Deaths, Murders and Lynchings " Abstracted from Lumpkin County, Georgia Newspapers Vol. 1 1873-1900 Courtesy Dahlonega Gold Museum

15. Meadows, John C. University of Georgia. Section on GEORGIA in the Encyclopedia Americana, Scholastic Library Publishing, Inc. Danbury, Connecticut 2004 p. 530

16. Smith, Sandi, Oklahoma State Historical Society op. cit. information on Dr. Emmett Starr, Rare Books, reprint 1921

17. FREEDOM'S CHAMPION newspaper, A. Martin Publisher, Atchison, Kansas Territory, October 2, 1858 p. 1

18. Lavender, David, BENT'S FORT, University of Nebraska Press, 1954, p. 278

and

Jessen, Kenneth, ESTES PARK, J. V. Publications Loveland, Colorado 1996, p. 1

19. Lavender, Ibid, p. 361

and

Broadhead, Edward, FORT PUEBLO, Pueblo County Historical Society, 1981 p. 7-8

20. Cox, J. Wendel, PhD, Senior Special Collections Librarian, Denver Public Library, correspondence

and

Manley, Joyce A., ARAPAHOE CITY TO FAIRMOUNT Johnson Publishing Co. Boulder 1989, p. 8

21. Line of descent for Lewis Ralston Sr. family Sandi, Smith ibid, Oklahoma State Historical Society, Oklahoma City, Oklahoma

INDEX